Evolution and Culture

Fyssen Foundation Series, published by Oxford University Press

Social Relationships and Cognitive Development
edited by Robert Hinde, Anne-Nelly Perret-Clermont, and Joan Stevenson-Hinde

Thought Without Language
edited by L. Weiskrantz

The Use of Tools by Human and Non-human Primates
edited by Arlette Berthelet and Jean Chavaillon

Origins of the Human Brain
edited by Jean-Pierre Changeux and Jean Chavaillon

Causal Cognition: A Multidisciplinary Debate
edited by Dan Sperber, David Premack, and Ann James Premack

Fyssen Foundation Series, published by The MIT Press

From Monkey Brain to Human Brain: A Fyssen Foundation Symposium
edited by Stanislas Deheane, Jean-Rene Duhamel, Marc D. Hauser, and Giacomo Rizzolatti

Evolution and Culture: A Fyssen Foundation Symposium
edited by Stephen C. Levinson and Pierre Jaisson

Evolution and Culture

A Fyssen Foundation Symposium

edited by Stephen C. Levinson and Pierre Jaisson

A Bradford Book
The MIT Press
Cambridge, Massachusetts
London, England

MIT Press books may be purchased at special quantity discounts for business or sales promotional use. For information, please email special_sales@mitpress.mit.edu or write to Special Sales Department, The MIT Press, 55 Hayward Street, Cambridge, MA 02142.

This book was set in Stone sans and Stone serif by SNP Best-set Typesetter Ltd., Hong Kong and was printed and bound in the United States of America.

Library of Congress Cataloging-in-Publication Data

Evolution and culture : a Fyssen Foundation symposium / edited by Stephen C. Levinson and Pierre Jaisson.
     p.   cm.
"A Bradford book."
Includes bibliographical references and index.
ISBN 0-262-12278-2 (alk. paper)—ISBN 0-262-62197-5 (pbk. : alk. paper)
1. Cognitive neuroscience—Congresses. 2. Human evolution—Congresses.
3. Brain—Evolution—Congresses. 4. Culture—Congresses. 5. Social evolution—Congresses.
I. Levinson, Stephen C. II. Jaisson, Pierre. III. Fyssen Foundation.

QP360.5.E966 2005
612.8′233—dc22

2005043348

10 9 8 7 6 5 4 3 2 1

Contents

Series Foreword

The aim of the Fyssen Foundation is to "encourage all forms of scientific inquiry into cognitive mechanisms, including thought and reasoning, that underly animal and human behavior; their biological and cultural bases, and phylogenetic and ontogenetic development."

The Foundation aims to support research that will lead to a more rigorous and precise approach to this fundamental domain, calling upon such disciplines as ethology, paleontology, archaeology, anthropology, logic, and the neurosciences.

The Foundation is named after its founder, Mr. A. H. Fyssen, a French businessman who was long interested in the scientific understanding of cognitive problems, and who was responsible for its endowment. Its headquarters are at 194, rue de Rivoli, 75001 Paris. The Foundation was recognized as a charitable institution by the French government by decree on March 20, 1979.

The Fyssen Foundation has developed a program to support research in the above-mentioned area. This program includes in particular:

- Postdoctoral study grants, for the training and support of scientists working in fields which coincide with the aims of the Foundation;
- Research grants to support scientists, in the field or in laboratories, who are working along lines of research corresponding to the objectives of the Foundation;
- The organization of its own symposia and publications on topics considered as important in fulfilling the aims of the Foundation;
- The regular publication of the *Fyssen Annals*, which include original articles in the fields supported by the Foundation;
- The Fyssen International Scientific Prize, given for a decisive contribution to the progress of knowledge in the fields of research supported by the Foundation. The prize is given each year to a scientist of international reputation.

The Foundation is administered by a Board of Directors consisting of scientists, lawyers, and financiers, and by an International Scientific Committee in charge of the scientific policy of the Foundation, of the launching of its programs, of the scientific evaluation of the applications, of the assessment of the work supported by the Foundation, and of the election of the recipient of the Fyssen International Scientific Prize.

Acknowledgments

This volume is loosely based on the proceedings of the Fyssen symposium, "Evolution and Culture," held in the Pavillon Henri IV at St. Germain en Laye, November 12–16, 1999. The contributors are most grateful to the Fyssen Foundation for the opportunity to discuss these issues in both depth and style. Madame Fyssen herself took a keen interest in the proceedings, as she did in all the doings of the foundation established by her husband, which supports the interdisciplinary study of the cognitive foundations of social life. This volume is dedicated to her memory.

The editors acknowledge and thank the staff of the foundation for their assistance in organizing the meeting, and Ms. Edith Sjoerdsma for her help in editing this volume.

Preface: It was not there in the Big Bang, but . . .

Pierre Jaisson

After long decades of mutual ignorance, the human sciences now seem to expect of the biological sciences, and particularly of physiology, neurocognition, and genetics, that they might explain the mechanisms of behavior and psychological states. It has also come to be expected of biology that it will provide, from the descriptive or experimental study of animal behavior, models that might point to useful conceptual and methodological tools. In this field, too, successes have been commensurate with the stakes, to the extent that sometimes the biological sciences have had a profound impact on the study of human psychology. This is notably true in the case of the conditioning model for animal psychophysiology or that of imprinting for ethology. However, even if such contributions are important, they only concern behavioral mechanisms—their "hows," their proximate causes. It is less obvious that biology has been able to address the question of "why." Indeed, a kind of sanitary cordon has long been firmly maintained around the subject of humankind, where there has been an emotional and ideological refusal to consider that human behavior might have any ultimate causality other than a cultural one. However, the theory of evolution, through its predictive and explanatory power as well as the spectacular progress achieved in molecular biology, seems at last to have broken through this barrier around the nature of humanity.

Had the social sciences known that sociality long preceded the hominids, whose main merit may have been simply to discover *culture*, they would have been called cultural sciences (which would indeed have spared us a number of debates). If it is too late to change the terminology, it must at least be admitted that the objective of the social sciences concerns only a very special kind of sociality (naturally, one dear to our hearts). However, after having read through a book such as this, it will become evident to the reader that at the beginning of the third millennium, the knowledge we now have makes the barrier between the biological and the human, between evolution and culture, ever more illogical. The ability to build a culture and to transmit it in a Lamarckian manner is the result of a long historical process through several major steps whose principal actor was natural selection.

In parallel with the interest in biological mechanisms, it has progressively become legitimate to ask why human psychology is the way it is. This in no way diminishes a proper interest in the functioning of our mental system, but rather the questioning helps us explore the evolutionary origin and the very nature of both our psychological functioning and our behavior, and thus may help us to better understand their finality, the situation of humankind in the animal kingdom, and finally, what is specifically human in our species. One can also imagine that the "why" question may provide (inter alia) some new insights into human behaviors that are deemed, by the norms of handbooks, schools, or cultures, quite pathological. Evolutionary biology is fully involved in explanations at that level also. Thus, a better understanding of the evolutionary significance of culture requires us first to know how to place it in the whole history of life on our planet.

Whereas no culture can exist without societies, nearly all animal societies *seem* to exist without culture. Sociobiology, the field that investigates the biological and evolutionary bases of social behaviors, studies hundreds of animal species which, like the human one, live in societies. Each of them shares with our own species a common ancestor that goes proportionally farther back the greater the phyletic distance between that species and ourselves. Thus, bees and humans share a common ancestor (which was neither insect nor vertebrate, and which was not social), and this common ancestor is more ancient than that which man has in common with the tyrannosaurus (certainly a reptile); it is older still than that shared by rats and humans (a more primitive mammal) and finally, even more distant from ourselves than the anthropoid primate whose heritage we share with the chimpanzee. If bees, tyrannosaurs, rats, chimpanzees, and humans are (or were) all social in their way, only the humans and the chimpanzees share a common ancestor recent enough to have been social.

As I pointed out earlier, man (let us say the genus *Homo*) did not in any way discover social life; he inherited it from an ancestor that he shares with many other primates, most likely with all monkeys. This comes down to saying that sociality is a life system that appeared independently several times during the course of evolution and considering just the five examples mentioned here, it can be stated that sociality appeared at least four times. In insects alone, sociobiologists count a minimum of fourteen independent occurrences of the phenomenon of sociality. Furthermore, not one of the ancestors common to all or part of those fourteen groups of social insects was already social. Despite everything, social life is rare in the animal kingdom. In insects, it concerns only 2 percent of the million or so species identified. In the midst of this ocean of solitary species, a rarity among rarities, human beings (the last living species of the genus *Homo*) have the unique characteristic of systematically setting up cultural societies that are sophisticated to a degree that is beyond the reach of other animals.

This does not mean, however, that social animals are mere automatons directly piloted by their genes.

As will be seen in this book, animal social cultures exist that are far less elaborate than those in man. In an even greater number of species, behavioral traditions can be transmitted from one generation to another by a nongenetic heredity, especially through parental behavior. These are passive transmissions, without motivation to learn or to be taught. A typical example has been described for the rat by Canadian physiologists[1] who demonstrated that highly manipulative mothers[2] induce in their offspring a lowering of stress when they are faced with an unknown environment— a property that becomes transmissible from the outset to the next generation. So the young adult females that went through such an experience in their youth present, as their mothers did a strong manipulative profile. It is thus possible, through adoptions, to reverse the fate of a rat. For example, an individual that would have become a stressed adult if it had been raised by its own only weakly manipulative mother will turn into an unstressed adult rat if it has been adopted by a highly manipulative mother. Resistance to stress can thus be nongenetically transmitted from generation to generation, even if the formal mechanism itself is obviously gene dependent. The transmission of the "stressed rat" or "unstressed rat" character is of the Lamarckian type, but it may be the only feature it has in common with an authentic culture in the human sense of that term. More surprisingly, some among the more recognized primatologists are reluctant to acknowledge the existence of culture in chimpanzees on the grounds of the absence of a pedagogical motive (the motivation to transmit and to receive knowledge). Readers of this book will find sufficient information to form their own opinions on the matter.

That we should find in animals few, or even no, premises or preludes to human culture should in no way prevent us from asking whether the social (noncultural) level that lies beneath the cultural level has an influence upon the latter, perhaps in the same way as the drift of continents is influenced by the underlying magma. A reflection upon the conditions under which sociality emerges could lead to conclusions that might cast some light upon the case of the human species, suggesting perhaps that the cultural exception has built itself upon a preexisting social base, modifying it in many ways. The qualitative jumps that in evolution occurred several times (sociality) or only once (culture) inevitably correspond to genetic aptitudes or predispositions that appeared, were selected, and have been maintained over time. This does not mean that experience does not have a role to play in the elaboration of individual social behavior or, even less, that the content of culture is determined by genes!

On the other hand, however, life in a cultural society that has universal rules (mutualism, value systems, hierarchies, religions, rites, superstitions, gathering and transmission of knowledge, etc.) implies that the framework of such rules should have appeared and been transmitted in the Darwinian manner, whereas the expression of such rules within each culture depends upon a Lamarckian transmission, that is to say, a nongenetic one. Animal sociobiology pointed out that "true" altruism, namely, altruism that is not preferentially oriented toward relatives, may only be selected and maintained when it is reciprocal.[3] The best strategy being to cheat rather than to avoid cheating, reciprocal altruism can occur only in particular conditions. One is the absence of cheaters because cheating is simply not in the program of the species. An example is the mutualistic supersocieties of ants in which several queens give birth to corresponding matrilines of workers who cooperate without preference for relatives.[4] The "complete confidence" which apparently rules an ant society might well explain why such supersocieties have appeared independently several times within this single family of Hymenoptera. Concerning the higher vertebrates, the story must be different, because cheating is widespread. Here, the higher cognitive level suggests that cheating has rather been negatively selected because of the cost imposed by sanctions against cheaters. Besides the cognitive capacities needed to punish cheaters (the recognition and remembering of the conduct of the cheater in the previous situation and the capacity to decide to inflict a sanction), the reciprocally altruistic vertebrate may live long enough to have a chance to punish a cheater in a reversed situation. Surprisingly, the effect of life-span on the probability of reciprocal altruism being selected in a vertebrate phylum is often underestimated. Although there already exist several evolutionary explanations for religion,[5] one may hypothesize that the widespread human belief about life after death was selected to postpone (and sometimes make endless) the time limit within which cheaters may expect to be punished. If so, this would explain why humans are highly reciprocal altruists and, possibly, why our ancestors won against other hominid competitors who did not have such an efficient mechanism for preventing cheating against one's own group. Of course, this is speculation. However, it gives an example of how evolutionary thinking may also contribute, among other things, to understanding human traits that are eminently cultural.

The discoveries of evolutionary biology applied to animal societies lead to an important observation. No species has passed directly from the solitary stage to the social. Indeed, in all cases social animals hail from ancestors who had a solitary parental life, thus a family structure both linked and limited to reproduction. Other things being equal, society is to culture what family is to society. *Family is the basic structure common to all societies.* Without it there would never have been either society or culture. This is no ideological or moral statement, but the mere observation of a zoologist. During reproduction time, many solitary species form labile family structures without attaining the level of a stable group that the social level implies. It is among these species that some animals and arthropods have sometimes crossed the Rubicon that leads to sociality. Most parental (familial) species never became, nor will ever become social;

like butterflies, they do not (and cannot) make up family groups. Family has always preceded society in the same manner as, with hominids, society long preceded culture. The theory of kin selection provides us with an explanation for this observation. The family brings together, in the same place and at the same time, individuals who share a large number of genes, thus considerably improving the likelihood that an altruistic act will indirectly have a genetic effect that is beneficial for its author. The evolutionist approach thus allows one to apprehend the family in the strict sense as a presocial structure that is in no way restricted to humans and that can be defined as a compulsory parental system organized between partners in reproduction and having the characteristic of gathering in time and space individuals who share a high number of genes. Society is, basically, a "superfamily" whose members maintain close ties beyond the period of reproduction. A perennial structure thus appears in the populations of a species. In the case of the human lineage, the step was made so long ago that humankind is not a good model to study if one is seeking to understand the emergence of societies. To try to understand sociality from the perspective of the human model would be rather like a botanist trying to describe the formation of a tree trunk by observing a blossom opening out at the tip of the finest twig of its branches. Better to stay at the level of the trunk itself!

Evolutionary biology is capable of explaining why social life, when it appears within a population of a species, generally gets the better of other populations of the same species that have stuck to the ancestral solitary parental formula. For understanding that, it is necessary to go back to the origins of sexuality; in other words, long before the appearance of the first family structure. Sexuality is a costly practice because of the random character of the survival of the young. However, it contains within itself a far higher adaptive advantage: that of increasing the genetic variability of descendants and thus conferring better resistance to disease. This probably justified the origin, as soon as sexuality came about, of a new evolutionary adventure: a strategy that seeks to form fewer female gametes but to endow them with better cytoplasmic reserves. Producing fewer ovules but producing them better and with a greater likelihood of success became possible provided that the production of spermatozoids was maintained at its original level, and provided that there was a correlative selection of new mechanisms that allowed them to be deposited close to the ovules. By beating the "genetic cards" at each generation, sexuality accelerated the emergence of interindividual differences, thus precipitating the rhythm of evolution and expanding biodiversity. There came a moment when, at different levels of the evolutionary tree, the improvement of the fertility rate could only progress through innovations in some domain other than that of physiological improvements, namely, behavior.

On the one hand, a set of coevolutionary processes affecting the sexual organs of higher plants and the behavior of animals (insects and certain birds or mammals) led to the latter becoming go-betweens in exchange for food (hence pollination). Parallel with this, animals started their own evolutionary process in which behavioral traits occupied center stage. The savings thus obtained were sometimes only slight and made no further evolutionary progress, as in the case of female butterflies which, instead of laying their eggs haphazardly in full flight, began to deposit them on the leaves of the plants on which the future caterpillars would feed. In many other groups, the result was made more spectacular through the emergence, on several occasions, of the parental strategy, and there we reach our starting point for social evolution: the family. We encounter this in fish (sea horses in which the males gather the young in an incubatory pouch); in amphibians (the male of the *Pipa* toad gathers its offspring in its dorsal pouches); in reptiles (the *Maiasaura*, meaning "good mother lizard," a large herbivorous dinosaur, would tend her brood of about fifteen young gathered in a nest 1 m deep by 2 m wide); in all birds; in all mammals; in some rare mites, spiders, and crustaceans; and finally in different insect taxa such as cockroaches, bees, ants, termites, and beetles. The adaptive value of the family is to be found in the fact that the offspring are protected more effectively and for a longer time and will only be left to fend for themselves much later. The investment made has a better guarantee of success and the parents disperse their genes more safely. Such is the *raison d'être* of the family, from an evolutionary point of view, of course. It is an aspect not to be neglected without forgetting that in turn the family was the starting point of the evolutionary adventure that led to sociality and then to culture.

Previously, before the family level, there were several other critical evolutionary steps that were also characterized by the pooling of entities of the previous level in a new, cooperative, emerging system. These cooperons[6] or major transitions[7] were successively the self-replicating macromolecules of life, the prokaryote cell, the eukaryote cell, the multicellular organism, the family, the societies, the societies of societies (ants and hominids), and finally culture. Each of these highly qualitative evolutionary steps emerged from one or another form of cooperation or synergy.[8] Does this common trait depend on a general law of the universe? We should have the answer once we become capable of really understanding the organization of extraterrestrial life. But that's another story . . .

Notes

1. D. Francis, J. Diorio, D. Liu, and M. J. Meaney, Nongenomic transmission across generations of maternal behavior and stress responses in the rat. *Science* 286(5) (1999).

2. These female rats display significantly more licking, grooming, and transporting behaviors toward their offspring.

3. R. Trivers, The evolution of reciprocal altruism. *Quarterly Review of Biology* 46 (1972): 35–57.

4. See for example, R. Blatrix and P. Jaisson, Absence of kin discrimination in a ponerine ant. *Animal Behavior* 64 (2002): 261–268.

5. P. Boyer, Religion explained. In *the evolutionary origins of religious thought* (Harper Collins, New York, 2002).

6. P. Jaisson, *La fourmi et le sociobiologiste* (Odile Jacob, Paris, 1993).

7. J. Maynard Smith and E. Szathmáry, *The major transitions in evolution* (W. H. Freeman, New York, 1995).

8. P. Corning, *Nature's magic: synergy in evolution and the fate of humankind* (Cambridge University Press, Cambridge, 2003).

1 Introduction: The Evolution of Culture in a Microcosm

Stephen C. Levinson

Evolutionary speculation constitutes a kind of metascience, which has the same intellectual fascination for some biologists that metaphysical speculation possessed for some medieval scholastics. It can be considered a relatively harmless habit, like eating peanuts, unless it assumes the form of an obsession; then it becomes a vice.

—R. V. Stanier, Some aspects of the biology of cells in H. Charles and B. Knight (eds.), *Organization and Control in Prokaryotic Cells*

As the quotation here suggests, this volume is full of the vice of speculation. Yet any student of the human condition can hardly avoid it. Somehow culture—or at least the culture-bearing ape—evolved. An evolutionary perspective on human culture, which is much less fashionable now than it was 70 or more years ago, seems inevitable, yet the social sciences actively resist it, allowing ill-informed conjectures from other sciences (which does little to increase the interest from the social sciences of course).

In this introductory chapter, I try to do two things: The first is to deal frontally, and speculatively, with what I take to be the "big questions" about the evolution of human culture. This may serve as a partial introduction to the more detailed explorations in other chapters in this volume. The second is to give the reader some grist for these speculative mills. I will argue that if we look at the details of any culture, it is quite clear that we need an evolutionary perspective to understand how such features could have arisen (note that such a perspective is quite consistent with other kinds of social science explanations). I will take as an example an island culture, which because of its relative isolation can serve as a microcosm in which to explore these issues.

The Big Questions

I take the following questions to express the fundamental issues that must be addressed in trying to develop a framework in which to think about the evolution of culture:

How Do We Embed the Phenomena of Culture within Evolutionary Theory?

We need of course to avoid the dichotomy the natives advance, with "culture" opposed to "nature," and find a way to see culture as just a part of nature, as evolutionary business as usual. The problem then is to determine what is the right framework in which to do this in order to situate culture within the scope of evolutionary mechanisms. What exactly are the properties of culture that make it special and out of the ordinary, and how are we to account for the fact that humans are obviously adapted to and for culture?

Why Did Culture Happen (more or less) Only Once? What Exactly Is the "X-Factor" in Our Lineage?

Exactly how often culture has evolved in the development of life on earth will depend of course on how it is defined. Finding culture among the apes is merely to find pale shadows of our kind of culture in our very own lineage, and thus to reassure us that there is indeed some evolutionary account for culture (rather than its being, for example, some accidental freak of nature). Even if we concede even paler shadows to some birds or cetaceans, there is no doubt that there is only one beast that has what Tomasello (1999 and chapter 10 in this volume) calls "ratchet culture"—the ability to build up ever more complex cultural and technological skills over generations. The question then is, what exactly lies behind this human ability, and what selected for it; that is, what is the X-factor and what kind of origin story can we concoct for it?

Let us take the questions in turn. There is a growing consensus, despite inbuilt resistance in the humanities and social sciences, that culture must be seen as part and parcel of the biosphere, and thus that there has to be some evolutionary story about humankind, not just as an ugly ape, but as a culture-bearing species with the ability and inclination to develop apparently limitless social and ideational complexity. However, there is absolutely no consensus about how to construct an explanatory framework for the origin of culture, and there are plenty of divergent strands of opinion. We need some framework that—without any magic "skyhooks" as Dennett (1995 and chapter 6 in this volume) has it—can provide the mechanisms by which the biological preconditions for culture could have evolved in the normal way that organisms evolve, and culture could have progressively raised the stakes, so that the biological preconditions were ratcheted ever upward. As Theodore Dobzhanzy (1962: 18) put it 40 years ago: "Human evolution cannot be understood as a purely biological process, nor can it be adequately described as a history of culture. There exists a feedback between biological and cultural processes."

The idea of a feedback relation between culture and genome is at the heart of what we can identify as the new synthesis, namely, "twin-track" theories of gene–culture evolution. There are various brands on the market (see e.g., Cavalli-Sforza and Feldman

1981; Lumsden and Wilson 1981; Boyd and Richerson 1985; Durham 1991), but they share the idea of a universal Darwinism, in which evolutionary theory embraces the study of all kinds of "replicators," where a replicator is any entity (such as a computer virus) that can copy itself by preserving information. Cultural entities, from items of technology to words, tunes, or fashions, can copy themselves through the actions of their hosts or "vehicles," suggesting a parallel between genes and "memes" or cultural replicators (Dawkins 1976, 1983: 109–112). Or, to look at it in a more conventional way, cultural entities can be transmitted by teaching and learning across generations, allowing "descent with modification" in Darwin's (1872: 3–10) succinct definition of evolution. The useful synthesis by Durham (1991) emphasizes the ideational nature of this cultural track, or line of descent, a point to which we will return. In these theories, both genetic and cultural tracks are seen as self-replicating strands, which share the fact that they are (1) informational, (2) partially independent, yet (3) potentially mutually influencing.

Both tracks are subject by hypothesis to universal Darwinian processes of selection. The interactions between the tracks produce a great upward spiral in "design space" (as Dennett has it), crucially with culture as part of the selecting environment, putting a premium on the underlying cognitive capacities that make the learning and production of cultural information possible. Figure 1.1 tries to represent this graphically, showing culture and genome spiralling up in design space over evolutionary time,

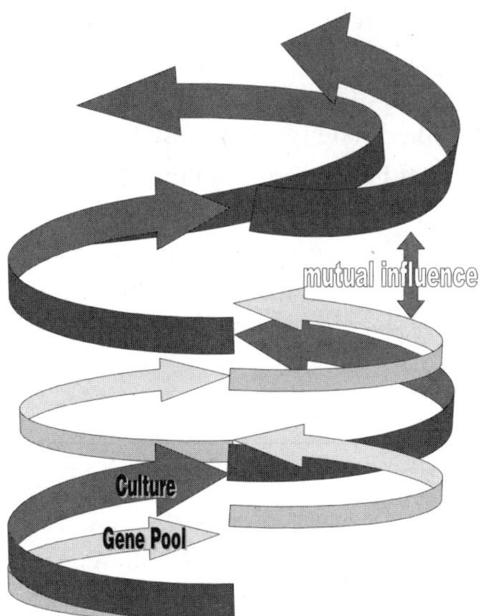

Figure 1.1
Twin-track evolution with feedback.

with an increasing independence as culture comes to have a life of its own, even being able to cushion the effects of natural selection on the genome.

There are a number of immediate challenges to this picture, which we should deal with right away. How can this close interaction work, the sceptic may ask, when the speed of adaptation in the biological track is severely restricted by the glacial pace of biological innovation (through mutation and reproduction over generations), while cultural evolution relies simply on ideational innovations that can jump the barriers of vertical (generational) transmission, and seems to be ever accelerating? However, there is an important existence proof that two separate strands of replicators that reproduce and adapt at quite different rates can nevertheless be deeply interlocked by mutual adaptation. I am referring of course to interspecific symbiosis. Take the classic case of the sycamore fig and the gall wasp (but see Combes in chapter 2 for an overview of different kinds of symbiotic relationships). The female wasp enters the fig inflorescence to deposit her eggs, and in doing so fertilizes the fig. She lays her eggs in the internal flowers; the males hatch first, fertilizing the infant female wasps while they are still embedded in the flowers, which have become galls. The males die and the females leave to repeat the cycle (Paracer 1986: 169–170). The fig relies absolutely on the wasp for fertilization, and the wasp only reproduces inside the fig flower; the two species are locked in an evolutionary embrace that neither can escape. Fig and wasp are constituted of separate tracks of DNA that can never intermingle; the fig provides the nuptial bed, the cradle, and the tomb for the wasps. They are obligatory symbionts, yet the male wasps live only days, whereas the fig tree can live for hundreds of years. In the same way, human biology and culture, though information channels with quite distinct pathways, different time trajectories, different vehicles, and different selection pressures, have become indissolubly linked. The human body is unlikely to survive without clothing, cooked food, tools, and other fundamental cultural replacements for teeth and claws. The parallels are spelled out in table 1.1.

The parallel takes us further. We know that some of the great leaps in evolution, "punctuations" if you will, have taken place in the special circumstances of twin-track

Table 1.1

Twin-track evolution: The parallel between symbiosis and gene: culture evolution

	Fig and wasp	Culture and human genome
Separate symbiotic tracks	Of DNA	Of ideas and practices and DNA
Context of life-span	(Male) wasp within fig	Individual within culture
Obligate symbionts	Fig depends on wasp, wasp depends on fig	Individuals depend on culture, culture depends on individuals
Differential speed of change	Many wasp generations to each fig generation	Culture changes faster than human genome

evolution, which allows a spiral of mutual adaptation in which the two strands can ultimately fuse into a single system. This is the story of the evolution of the eukaryotic cell. [The mitochondria and chloroplasts that occur in eukaryotic cells seem to have originated as separate organisms that took up residence inside other cells; see Margulis (1981) and Combes in chapter 2]. In a somewhat similar way, the interaction between culture and genome in humans has produced an extraordinary symbiotic hybrid, yielding a quantum leap in adaptational flexibility, which allows humans to exist in every niche on the planet and beyond. The process appears to have been initially gradual, but then to have accelerated (see Foley in chapter 3 in this volume).

A second immediate challenge to theories of twin-track evolution is the uncertain nature of the feedback mechanisms between the two tracks. Even in biological mutualism, such feedback systems are still somewhat unclear, but those involved in gene–culture evolution seem entirely obscure. There have to be potential feedback mechanisms in both directions: from genome to culture, and culture to genome. The genome to culture route is easier; it is clear that biological capacities can facilitate or make possible cultural adaptations. Thus the anatomy of the hand and its motor control were the prerequisite to tool making, or the human vocal apparatus and its neural underpinnings were the foundation for producing language. Culturally variant technologies and distinct languages rest upon a shared biological foundation. Thus during the course of human evolution, neurophysiological adaptations (e.g., in cognitive capacities and motor skills) would have afforded new cultural forms. However, the feedback loop from culture to genome is altogether harder. How could the growing use of tools, for example, have possibly influenced the underlying neurophysiology and anatomy that supported it? How, in short, can one explain Lamarckian effects while avoiding the Lamarckian fallacy?

There are in fact a number of perfectly plausible mechanisms by which culture can have feedback effects on the genome. Here, for example, are half a dozen candidates:

1. Natural selection in a cultural environment Selection operates on the phenotype through the environment. That environment can be partly constructed by the organism itself, which is involved in a process of niche construction (Laland et al. 2000). Thus the selecting environment can be cultural. Take fire. The use of fire softens food and increases its recoverable nutritive value, while rendering plant toxins and bacteria in flesh relatively harmless. It probably lies behind the progressive reduction of dentition that we associate with the development of *Homo sapiens* from *Homo erectus*, and perhaps also changes in the gut that are not visible in the fossil record.

2. The Baldwin effect Behavioral changes can feed back into the genome by exposing the organism to new environmental pressures, e.g., through the colonization of new ecological niches, where natural selection can eventually "fix" a surrogate of the behavioral adaptation in the genome (Baldwin 1896; Deacon 1997: 322–323). Take

clothing. With it, early humans could move into much colder environments, where natural selection would gradually favor a less gracile, less African body shape, one better adapted to the same circumstances that the clothing was adapted to. Thus the Inuit (Eskimo) peoples have evolved special physiological adaptations to extreme cold (such as vasodilation in the limbs, the ability to produce glucose from meat instead of from carbohydrates), but it was only the technology of creating clothing from skins that made inhabiting the arctic possible in the first place.

3. Group selection   In the normal case, natural selection operates on individuals. Among humans, however, where culture-bearing units compete for resources, if a group can act as a single whole (e.g., in conflict), it may survive or perish as a whole (Wilson and Sober 1994). Cultural or technological superiority can then force a less well-equipped group into extinction (see Boehm 1996 and chapter 4 in this volume, and Boyd and Richerson in chapter 5, for limitations on the mechanisms). This is a likely scenario for the Neanderthals of northern Europe, and recent research (Krings et al. 1997) suggests that our genes have triumphed without sharing theirs.

4. Kin selection in culture-bearing kindreds   Group selection can be related to the better-understood mechanisms of kin selection, in which altruistic behavior by one individual on another's behalf can benefit the donor if there is sufficient biological relatedness between them. Small groups, such as hunter-gatherer bands, are kindreds; they are typically both culture-bearing units and descendants of a common ancestor. The mechanisms of kin selection will thus tend to keep the minimal culture-bearing unit afloat (see Boyd and Richerson, chapter 5).

5. Sexual selection   Darwin (1871: ch. 8) considered that many aspects of human morphology could be attributed to sexual selection, that is, the choice by females of their mates. If a culture sets up boundary conditions for reproduction (e.g., the payment of a bride price or success in economic or political arenas), it translates the biological foundations for those social skills into reproductive success, thus ultimately fixing those skills in the genome. Humphrey (1976) introduced the idea that the most demanding aspects of the environment for primates are their own typically complex social systems, and thus that primate intelligence can be ascribed to the mental gymnastics required to survive and reproduce in there systems. A quite plausible story for the evolution of the biological foundations for language can be told in terms of sexual selection (Deacon 1997).

6. "Auto-domestication"   Every dog reminds us of the power of selective artificial breeding, even over a relatively few generations. In this case, humans are the breeders, the dogs our slaves. Of course we do selective breeding in our own species, as in strategic infanticide and infant neglect, in planned marriages and the harems of despots. We also selectively invest in our young, differentially affecting the reproductive success of our offspring. Darwin (1871: ch. 2) speculated about these effects in the planned marriages of Prussian soldiers and in Spartan infanticide, but noted that

there were distinct limits to the extent to which humans have applied breeding techniques to their own kind, even in slavery. Still, unpalatable though it may be, some degree of auto-domestication is probably observable in every society, and may have been rife in human history (see e.g., Voland 1998; Beise and Voland 2002).

Such a list of mechanisms is bound to be controversial, on account of the wars between the ultra Darwinists (such as Williams 1966; Dawkins 1976; and Dennett 1995) and the soft Darwinists (such as Gould and Lewontin 1979 and Rose 1997), who don't like Darwin's dangerous idea neat, without a dash of seltzer (see Sterelny 2001 for a dispassionate assessment). The difference is that the ultra Darwinists emphasize one level of replicator, the selfish gene, and strict adaptationism, while the soft Darwinists emphasize the whole hierarchy from gene to organism to supraorganism (e.g., distinct species in symbiotic association), and embrace the idea of other mechanisms, such as symbiosis, group selection, niche selection, "spandrels" and exaptation, and even wild chance. When considering the evolution of culture, we are dealing with a rare (even unique) event, closest in kind to symbiosis or mutualism, and it perhaps makes sense to entertain the broadest array of possible mechanisms. In any case, it will suffice for the current argument that at least some such feedback processes—by which cultural attributes might have effects on the genome—can be presumed to exist. Certainly it is hard to envisage how our species could possibly have evolved the way it has without adaptation both to and for culture—naked and defenseless, slow yet terrestrial, with a gut and dentition requiring prepared food, with nearly a third of the natural life-span spent in dependent childhood.[1]

Perhaps the most powerful argument is, again, an existence proof. Durham (1991: 226–285) develops a careful argument from earlier work that distinct food-preparation techniques have resulted in microevolutionary adaptations in human groups. Thus human groups differ in their ability to digest milk in adulthood. In most humans, the production of the lactose-digesting enzyme, lactase, shuts down after weaning, and many individuals then exhibit severe side effects from drinking milk. However, in northern European dairying populations, and equally among milk-drinking African pastoralists, most individuals can absorb lactose throughout adulthood. There is an interesting intermediate group that contains both lactose absorbers and nonlactose absorbers, such as peoples on the Mediterranean fringe. Here dairying takes a different cultural form; fresh milk is first soured and turned into yoghurt or cheese, where bacteria digest the lactose. The fit between cultural practice and the distribution of relevant genes in the populations is precise enough to strongly support the causal relation, and alternative hypotheses either fail detailed analysis or are compatible with the role of agricultural practice as the dominant casual factor. In addition, the kind of time scale involved, in thousands of years, is sufficient for this kind of feedback to have taken place.[2] If we can demonstrate this kind of microevolution under cultural feedback in perhaps as little as forty generations, there is no reason to doubt that similar processes have taken place in many other cases.

Other examples of the effects of cultural adaptation on the genotype can be seen in the differential prevalence of type II diabetes and hypertension across cultures. These have sources in metabolic thriftiness and salt conservation, respectively—adaptations to insecure, difficult, preagricultural conditions. Indigenous populations that converted to agriculture and trade suffer a much higher incidence of these diseases than Europeans, who have adapted to "the unconscious domestication of humans by agriculture" (Diamond 2002: 707). Durham (1991: 103–153) also documents a similar interaction between agricultural practice, malaria, and sickle-cell anemia in West Africa.

A third challenge to theories of twin-track coevolution is a rejection of the parallel between information in the genome and information in the cultural track. Cultures, the challenge goes (see e.g., Midgely 2001), are not built out of replicators or "memes" to be likened to genes, as Dawkins (1976) suggested. Indeed, the idea of memes—without clear ideas about the larger structures built out of them—has incurred the ire of social theorists, who do not like to see the highly integrated institutions of social life treated as unstructured masses of memes.[3] Sociocultural systems are systems, not heaps of transmissible traits. The parts of systems cannot be replaced by random objects unless the objects happen to fulfill the same function as the parts (think of a spare part in a car engine). So how seriously should we take the theory of memes? Leaving aside its sufficiency for a theory of culture, there are fundamental internal unclarities about memetics, and serious deviations from the genetic model, as the following list suggests (see Dennett 1995:352–360 and chapter 6; Midgley 2001):

• DNA has only syntax, while memes have semantics; copying of genes is mechanical, copying of memes is intentional and pays attention to the semantics. Dennett and Sperber (chapters 6 and 7 in this volume) explore the implications of the fact that people reproduce what they take to be the intentions behind actions (hence the importance of a theory of mind, which is extensively discussed in this volume). While Dennett suggests that this can be thought of as just a higher-level editing procedure, Sperber holds that this dooms memetics.

• Genes do not cross lineages, memes routinely do; there is horizontal as well as vertical transmission. It is notoriously difficult to tell diffused traits from lineally inherited ones in culture. In anthropology in the 1920s, this was a major contention, and even today there are neighboring languages where the experts can't agree whether the languages are related by descent or by convergence through borrowing.

• Memes by descent versus convergent memes: How can one distinguish inherited memes from parallel invention, especially since part of an idea might be inherited and the rest developed independently? Consider also that evolutionary psychologists like to think that many alleged memes are just phenotypic expressions of genes.

- Memes blend in the very process of replication, genes do not.
- The fidelity of reproduction: DNA replication is nearly perfect but memetic reproduction is mutuation-prone. Indeed the transformation of memes in reproduction might be one of the most important properties of cultural facts, suggesting an alternative model (see Sperber in chapter 7).
- Memes have to be isolates, and thus independent units. However, cultures (as mentioned) are structured assemblages of ideas, and there is no obvious basic unit. Take a kinship system. Is the meme the notion of "mother's brother," "uncle," "collateral kin," or indeed the whole structured system of kin terms, or is it the associated behaviors or the inheritance customs built on that system, or what exactly?

In short, memes do not really look like genes. They lack Mendelian principles of segregation and assortment, veridical copying, and vertical descent, and they pass through the active filters and recombinations of the mind, not automatic splicing and editing procedures. There are many further basic issues; for example, if genes are selected via their phenotypic effects on their vehicle (the organism), what exactly are memes, meme vehicles, meme phenotypes, etc? Much more work would be needed before we could be said to have a serious theory of memetics (see Aunger 2000; Laland and Brown 2002). Sperber (chapter 7) suggests that a proper analysis of cultural reproduction is going to involve recognition of mental representations that play causal roles in action sequences. The whole reconstruction of social science along naturalistic lines is required, in a direction more like medical epidemiology than the operation of natural selection on DNA.

It is therefore important to see that the theory of twin-track coevolution doesn't depend on the meme. The units of transmission and the modes of cultural reproduction may be quite varied, yet the reproduction of culture across time and space is indisputable. We can note that cultural phenomena are both vertically inherited and horizontally borrowed, and that the units transmitted (or, rather, actively learned and used) can be individual traits or whole systems (I will give the example of number systems later, but there are also cases of whole languages being borrowed to replace an inherited one). We can also note that what can be passed on can be a situated practice (such as how to make a canoe), residing as much in honed motor skills as in mental templates. Consider piano playing. The piano itself is the product of a long cultural heritage of making stringed instruments, but the object does not by itself convey Beethoven concertos or even the art of the pianist. There is a whole cultural complex surrounding the piano, combining craftsmanship in wood, motor skills, long musical training, notions of musical intervals, the conventions of musical notation, the practice of musical performance, and so forth. As this makes clear, treating culture as a set of purely mental abstractions in order to set up the gene–meme analogy is to artificially divorce the concepts from the behavioral practices that support them and through which they are learned.[4] In the second part of this chapter, I will return to a consideration of the essential properties of culture that need to be taken into account by any theory of cultural reproduction.

Let us now turn to the second fundamental question raised earlier, namely, if culture is such a good thing, why did it happen (more or less) only once in the history of life on earth? And why did it happen in our lineage, and not in lineage after lineage? First, a few caveats are in order. As mentioned earlier, the rarity of the phenomenon depends somewhat on what you mean by "culture" (a subject fraught with disagreement; see e.g., Fox and King 2002). Even if we take it to include any kind of behavior transmitted by learning across generations, which thus marks off one group of the same species from another, culture is still going to be strikingly rare, with some few examples from, e.g., the oscine birds (Hauser 1997: 273–300), whales (Rendell and Whitehead 2001), and chimpanzees (Whiten et al. 1999; McGrew 1992). These exceptions hardly alter the picture, given the limited and specific domains involved, because they do not exhibit the property of indefinite accumulation of innovations that is the signal mark of human culture (what Tomasello calls "ratchet culture"). Nevertheless, they may be crucial for understanding the full-blown primate version. Every trace of a parallel to human culture is to be welcomed, since the discontinuity of human culture from everything else to be found in nature is an essential problem for evolutionary theory. A second major caveat is that we, of course, are not the only species to have had highly developed cultures. However frugal you are with the appellation of "culture," there will be at least a dozen hominids (many of whom are not our ancestors) that had it! Thus the Neanderthals, who now seem clearly to have belonged to a distinct lineage (Krings et al. 1997), had all the trappings of hafted tools, clothing, control of fire, funerary rites, and so on. We share with all these rival hominids a relatively recent ancestor (within the last six million years or so).

So what was the X-factor, the magic ingredient for culture? Of course we are not interested in all the preconditions of culture—that would take us all the way back through the great chain of being. What we are interested in is the added something that takes us out of the general run of mammals, or indeed of other highly social organisms like the ants. A good way to get a grip on the X-factor is to look at our nearest relatives and ask, do they have it, and if not, why not? A good case can be made for chimpanzee culture (Whiten et al. 1999), although the differences among groups may have an as-yet unascertained ecological basis. In this volume, though, the predominant view is that chimpanzees do not make the grade. Premack, Hauser, and Tomasello all concur in that opinion, although Hauser and Tomasello offer different reasons why they do not.

There are a great many behavioral and functional candidates for the X-factor in the human lineage, and these preoccupy other authors in this volume. For example, Dunbar argues for a special role for advanced communication systems in holding large groups together. Tomasello emphasizes the special role of social and cultural learning and the theory of mind (or "mind reading") that underlies it. Boehm and Hauser identify the nature of morality and inhibition, while Foley ridicules the idea of any single X-factor, emphasizing instead multiple dissociated factors such as planning, technology, learning, and language. One thing all these functional candidates have in common is that they would have driven the evolution of higher mental capacities, and thus the development of a larger brain.

Cognitive Candidates for the X-Factor

The brain seems to have been, along with the hand,[5] the central locus of gene–culture coevolution—the main organ on which selection pressures favoring culture must have worked. The human brain is about three times larger than might be expected for a primate of our size, and the development of this greater capacity is roughly correlated over a period of two million years with increasing cultural complexity, as measured by tool use (see Foley in chapter 3). What is so great about a large brain? Ants can calculate complex navigational paths without one, busily checking solar ephemerides and doing trigonometry (Gallistel 1990), and Gallistel, Gelman, and Cordes in chapter 12 try to persuade us that when it comes to math, the main human advantage is being able to talk about concepts that exist antecedently in primate cognition. In general, there is no consensus across the cognitive sciences about what makes our large brain such a special computing device. One assumption (lying behind the careful comparative study of brain size in, e.g., the chapters by Foley and Dunbar) seems to have been that what is crucial is largely a matter of the sheer quantity of RAM as it were; that is, a relatively undifferentiated neocortex. The specifics then depend on learning. Universal Darwinism can be applied to the developmental trajectory of the neocortex, and one can look at the whole gigantic wiring of synaptic connections as a selective process (Changeaux 1985; Edelman 1987; Singer in chapter 9), with distinct kinds of selection—natural selection over deep time, developmental selection during early ontogeny and brain maturation, and selection over the life-span in response to environmental and cultural pressures. This approach suggests that the magic ingredient is not so much a specific genetically determined brain component (running a single "native" machine code as it were) giving us Culture (with a capital C) in the singular, but rather a highly adaptable computing device that can run any number of high-level programs, giving us cultures in the plural. In chapter 9 Singer introduces an interesting twist to this argument. The evolutionarily new areas of the brain, although composed of much the same tissue, are secondary association areas, which take their input from older primary sensory areas. The new areas seem to be especially dedicated to multimodal metarepresentations, realized as massive assemblies of cells that bind lower-level representations through synchronicity of firing. It is these metarepresentations, indefinitely stacked, that make possible a theory of mind. Thus more of the same brain tissue can have emergent properties, in this case the particular properties (as Tomasello and Dunbar argue) essential to culture.

Instead of fixating on encephalization (or sheer brain size relative to body weight), there is another way of looking at the brain, namely, as a ramshackle collection of ancient modules, or specialized processing units, for some of which we may be able to tell good adaptationist stories. The rival assumption, then, is that the phylogenetic additions to the human brain are qualitative, giving us highly specialized brain tissue, as most clearly exemplified by the language areas of the brain. These additions may have provided us with a range of functionally specific "modules," adaptations targeted by natural selection, perhaps for many areas of activity, including technology and Culture (with a capital C, i.e., the underlying learning capacities and motivational structure). In general, most students of comparative neuroscience agree that it's not the sheer size of the human brain that should be the focus of evolutionary speculation, but rather those areas that show the greatest relative growth (see Dunbar and Singer in chapters 8 and 9).

Deacon (1997) argues that language is the key functional adaptation, and there is much in this volume to support that idea. Hauser (chapter 11), however, emphasizes the frontal lobes, which are strongly associated with inhibition. Without the inhibition of immediate reflexes, physical or mental, there is little possibility of maintaining a single train of thought, let alone the extensive planning of future action, or any kind of morality. In fact, Hauser shows that lack of inhibition, like alcohol, masks the considerable reasoning powers of many primates. Other candidates for uniquely human cognitive skills include mathematics. However, in chapter 12, Gallistel, Gelman, and Cordes take a close look at numerosity skills across species and come to the conclusion that human mathematical skills are based on a phylogenetically ancient system of estimating quantity that is transformed only through language.

This modular perspective, at least as defining human cognitive skills, can thus be taken much too far. The doctrine that human cognitive abilities are all innately coded in the brain dominates the cognitive sciences, and this has been extended to what are taken to be the major cognitive features of culture (see e.g., Barkow et al. 1992; Jackendoff 1992; Talmy 2000, vol. II: ch. 5, Plotkin 1997; and responses in Rose and Rose 2001; Brown 2002). The reasoning is that the mind is the fundamental filter for possible cultures. Every meme that cannot be learned or processed by the mind dies an instant death (a point that is true as far as it goes).

The next (and completely unwarranted) assumption is that there is not much else to culture; to understand the mental filter is to understand the biology/culture interface, or as Plotkin (1997: 253) puts it, "a theory of culture is first and foremost a psychological theory." Cultural variants are just meaningless variation—noise in the system. Such a stance fits with a number of strands of ultrareductionist thinking in current thought, in evolutionary psychology, and in sociobiology, and with the extreme forms of nativism in the cognitive sciences. However, it mistakes a precondition for culture for the phenomenon itself, which is not a psychological phenomenon but a historical one in an ecological context, and not a property of an individual, but a property of a population with elaborate divisions of labor and knowledge.

The problem with this kind of nativist reductionism, in which there is nothing in culture that is not essentially in the organism, is that it has lost sight of the central phenomenon. It is the variability of culture that is responsible for its adaptive value, by extending the phenotypic range of the genotype. The quite extraordinary thing is that in prehistoric times we already inhabited lands with permafrost and scarcely any vegetation on the one hand, and lands with unremitting heat and scarcely any water on the other, and about 30,000 years ago we had crossed hundreds of miles of open water by boat (see later discussion). The cultural basis for this radiation is advanced technology—of clothing, boats, food processing, desert navigation, and so forth. If there was only one form of culture, for example, only one form of kinship, only one kind of political system, only one set of tools, only one type of religion, we would all happily be ultrareductionists talking of instinct instead of culture (and we would all still be in Africa). That is not the phenomenon we are trying to explain.

Take language. The most astounding fact about language is that it is variable in both form and content. We are the only species in nature's vast spectrum of organisms with a communication system that varies in both form and meaning.[6] We can even change its modality from the vocal-auditory channel to the manual-visual one, as in the natural sign languages of the deaf (a point I will return to), or in the written modality for that matter. No other animal can do that. Theoretical linguistics has, under Chomsky's guidance, been preoccupied with discovering the underlying architectural commonalities in language, dubbed universal grammar. The term has suggested to many nonlinguists that there is really only one language, with superficial clothing of different kinds (different sounds, for example). Nothing could be further from the truth. Languages vary in fundamental ways; for example, they may have as few as a dozen distinctive sounds or as many as a dozen dozen (141 is the record); they may or may not have morphology (affixes on words, such as the plural on "pen-s"), they may not make familiar word-class distinctions (as in English nouns versus verbs), they may or may not have a fixed word order, and so forth.[7] Languages are so diverse that establishing even a short list of universals (in the sense that all languages have them) has proven frustrating; they tend to be trivial predictions of the sort that all languages have at least one vowel (otherwise you could hardly hear them!). Most empirical universals are conditional predictions of the sort "If a language has property X then it probably has property Y," nearly always with attested exceptions. And we still have more than 90 percent of languages to look at!

Ultrareductionists have tried to treat language, just like the communicative systems of other species, as an instinct (see e.g., Pinker 1994). They have claimed that language is an essentially universal medium for broadcasting universal thoughts. This is absurd; a particular language reconfigures our thoughts. So much of our cultural heritage is encapsulated in cultural concepts packaged into words. I don't expect the Rossel islanders of Papua New Guinea (see the next section) to comprehend the notion of a sonata or calculus, anymore than I find it easy to understand their concept of *ngm:aa* ("shell coin one denomination lower given as security for the loan of a shell coin one denomination higher, in the series of shell-coins called *ndapî*") or *chimi* ("two persons A and B such that A stands to B in the kinship relation *kênê* and B stands to A in the kinship relation *chênê*"). Elsewhere I have called the doctrine that holds that not only formal operations, but also the very content of our thoughts are determined by our genes, simple nativism (see Levinson 2000). What is right about simple nativism is that it insists on the prestructuring of our mental abilities. What is wrong about it is that it minimizes or ignores the role of ontogeny and learning, and minimizes the very stuff of our evolutionary success, namely, the cultural variation that is our special system for rapid adaptation to differing environments. Culture is a way of generating phenotypic variants far broader than a strongly canalized expression of the genotype alone can manage.

If the mental X-factor is not a set of prefabricated systems with standardized output, what is it exactly? Quite clearly it is a set of learning mechanisms that can accept a broad spectrum of input, but can output a narrow band of acceptable behavior that is in line with the very specific local input. Language again reminds us of the fundamentals. The sound systems of languages vary enormously. Rotokas has just eleven distinctive sounds (and five vowels), while Rossel Island language (Yélî Dnye) in the same Island Melanesia geographic region has ninety (and thirty-three vowels), some of them unique to just that language. This is a huge difference in auditory discrimination, so the learning mechanism must tolerate that broad spectrum of possibilities.

Human infants have special cognitive abilities that are built for exactly this cultural variation. For example, in the realm of vowel sounds, infants of just 6 months have been shown to restructure their auditory space according to the local language; the space becomes systematically and irreversibly distorted, so that sounds that are acoustically equidistant will now become assigned to the same or different categories along language-specific lines (Kuhl and Meltzoff 1997). The end result is a range of spectacular biases in our auditory perception, which make adults unable to even hear the difference between sounds that are fundamentally distinct in some other language. Thus the initial perception system ends up systematically skewed. Exposure to a specific language rebuilds our perceptual acuities, and it does so at such an early age that it seems inescapable that the system is built for handling diversity. [We know that other primates exhibit a similar tendency to hear sounds as belonging to categories, but not this ability to distort the acoustic space through learning; see Kuhl (1991), Hauser (1997: 324)].

This example may serve as a token of the special kind of cognitive ability that is required for a culture-bearing species. In this case the infant needs to know in advance that speech sounds are important. It needs to presume that there are significant local categories to be discerned using complex statistical pattern matching, and then it needs to learn to ignore some sound distinctions while acquiring heightened perception of others, thus distorting acoustic space in line with the input. We may expect that some of these cognitive underpinnings for learning cultural variants are highly modality-specific, like our vowel example, while others may be general learning capacities. Culture-acquiring children cannot be preprogrammed to shoot accurately with a bow anymore than to play the piano (both are local cultural objects), but they can be built to expect highly precise motor routines in hierarchically organized schema for specific cultural purposes. Obviously there are many other preconditions for culture, including the motivational structure, metarepresentational abilities (chapter 9), special memory abilities, and the ability to inhibit antisocial urges (chapter 11). However, these alone will not give you culture; for that, you need the special ability to know what kind of pattern to look for and to identify the local variant in a broad spectrum of possibilities. It is for this reason that "mind-reading" abilities are correctly emphasized by Tomasello, Singer, Dunbar, and other authors in this volume.

If simple nativism (reductionist evolutionary psychology) is an explanatory dead end, the reason is that it has lost sight of the explicandum, the variable end product that is the whole advantage of the human mode of adaptation. We can grant that human cognition is highly structured, but culture is not the projection of that structure alone. The theory of coevolution offers us a much better way to think about human cognitive abilities—as developing in a space with distinct attractors, both cognitive and cultural. Take the words for colors in different languages. Some languages have only two basic color words, some have eleven or more (not counting specialist terms or kinds of basic colors; see Berlin and Kay 1969; Hardin and Maffi 1997). There is a correlation with culture; cultures with dyes and weaving, paints and painted decoration have more developed color terminologies than those that do not. The limiting case is a language like that spoken on Rossel Island (see later discussion) where there is no technology of color and only the rudiments of a system of color terms (Levinson 2000). However, as a culture acquires interest in color words, there is a distinct order in which color words are "fractionated" out of more global cover terms, so that soon simple white, red and yellow, and green or an amalgam of green and blue are labeled. The order seems to reflect perceptual salience, influenced by cultural preoccupations (e.g., Mesoamerican obsessions with turquoise and jade may have attracted solutions in the direction of a green and blue amalgam). The outcome is a balance between perceptual salience and cultural interest.

Generalizing the model, the idea is that the architectural complexity of any human cognitive ability can be apportioned between two sources: innate predisposition on the one hand and cultural and experiential input on the other. Native predisposition may be less in the way of innate ideas (representational nativism) and more in the way of constraints on information processing, owing to the structure of the brain and perceptual organs (architectural nativism; see Elman et al. 1996). These cognitive constraints have evolved hand-in-hand with culture, and they bias cultural transmission. By bias I mean they weight the probability of exact replication, or, if one prefers (see chapter 7), affect the chances of transformation.

The advantages of the coevolutionary account of human abilities is that for most absolute universals it is possible to come up with cultural counterexamples; for example, a culture in which brothers marry sisters (Ptolemaic Egypt), or languages that do not have well-established color terms or have unique sounds that are not made in any other language (Rossel Island), or cultures in which different spatial coordinate systems are used in everyday cognition (as in Guugu Yimithirr; see Levinson 2003), and so on. However, universals of a statistical kind are much easier to find, implying systematic biases; there are historical, social, and ecological conditions that set functional constraints, and there are cognitive limits on reproducible ideas and practices.

So where are we? The X-factor is a set of cognitive adaptations for culture. They include necessary preconditions, such as the mind-reading abilities emphasized by Singer and Tomasello and the inhibition emphasized by Hauser. However, they cannot include precise instructions for the contents of a culture in the way that the evolutionary psychologists propose; cultures are just too variable for that. Rather, they must include specializations for tuning in to local cultural patterns, as illustrated by the case of language where long before they understand a word, infants are already shaping a soundscape for the language they will learn. This allows rich local adaptations to be preserved in the cultural environment in which the child grows up.

Why do no other species use the same trick of displacing highly detailed adaptation into a cultural mode? If culture is that good a trick, why did we have to wait three billion years for our species to come along and exploit it? Here we have to admit that current explanations are rather feeble. Biological anthropologists stress that in metabolic terms the brain is ultraexpensive tissue to maintain (Aiello and Wheeler 1995), and large crania make for dangerous births. But this only suggests that a monkey or dolphin with a bit more gustatory effort might have developed cultures, which they haven't. Perhaps no explanation is necessary. Is culture simply in such a distant nook in Dennett's (1995) design space of possible adaptations that the chance of any other species traveling there is infinitesimally small? Or perhaps there is more transmission of learned information going on in other social species than we can currently discern. None of this is satisfying, but the answers must wait on the comparative biologists.

Some Crucial Properties of Culture—Reminders from an Island Culture

For most chapters in this volume, culture is the explicandum. What exactly do we need to account for? Here, as reminders, are some crucial properties of culture, which any theory of cultural origins must take into account:

System complexity without a single designer   Just like organisms, social and cultural systems display intricate designs beyond a level that could be achieved by any individual designer, even if there was one. Consider, for example, languages, kin classification systems, rituals, large-scale irrigation systems, or building methods. It is systems rather than traits like memes that need an evolutionary account.

Multiplicity and variation   There are thousands of distinctive cultures (taking languages as a gross indicator, there are on the order of 6000 to 8000 today). The variation across cultures consists both in distinctive elements and in distinctive arrangements of elements (consider, illustrating again with language, distinctive phonemes and distinctive arrangements of phonemes into possible word forms). Until recently at least, geographic and social separation correlated with cultural difference; cultures left to their own devices diverge. There is also variation within cultures, which often provides the source of innovation and change.

Vertical transmission and cladistic character   Cultures derive partially by "descent with modification" (in Darwin's one-liner defining evolution). Thus the cultures of far Oceania are all derived from the Lapita culture of the first Austronesians, the first human colonists of the area. Many of the basic adaptations, such as crops, outrigger canoes, and stilt houses are still shared from this ancestral culture of c. 5000 years ago. Through vertical transmission, cultural phenomena can become extraordinarily stable. The bifacial hand axe holds the record as a cultural object that remained essentially unchanged for a million years, but there are many cultural ideas and objects of more modest antiquity, for example, languages with 3000 years of continuous written history (Chinese, Tamil), Egyptian statuary conventions with 3000 years of continuity, elements of architectural style like the Corinthian column with 2000 years of nearly continuous use, the spoked wheel with an ancestry of at least 4000 years, the mason's round mallet with 5000 attested years of use, the alphabet with nearly 3000 years of antiquity, and so forth.

Horizontal transmission and cultural diffusion   In addition to vertical transmission, cultural ideas and techniques are borrowed. Consider the technology of warfare, Creole languages, or song styles. However, there are also striking examples of parallel invention, similar to parallel evolution in biology (as in the placental versus marsupial analogues of moles, wolves, mice, etc.). These suggest that the design space has some tight intrinsic limitations.

Cultural phenomena are cumulative   They can embody the wisdom of generations of experience, e.g., how to process poisonous plants, which fish to avoid, what to do in the case of rare environmental catastrophes. This is one of the crucial adaptive advantages that humans have over other creatures. We inherit the results of millennia of experimentation without any of the costs or dangers.

Cultural phenomena are not always adaptive   Cultures include large proportions of elements that appear functionless, sometimes even deleterious, to both cultural preservation and biological success. An evolutionary approach can help us to understand how such features are nevertheless propagated.

Culture, group selection, and the feedback to biology   Cultural groups can act as wholes, taking collective decisions upon which the biological survival of the whole group depends, as in war or response to environmental catastrophe. They can also be extinguished by ethnocide and by enculturation into larger groups. Such processes can be observed in real time. Even processes of microadaptation or coevolution between a culture and the gene pool can be directly observed.

Such a list immediately suggests the necessity of a Darwinian account. Cultures have design without a designer, evolving by the selection of existing elements of variation in accord with functional requirements, speciating in new ecological niches, with elements transmitted over scores of generations. On the other hand, some elements of culture are more like parasites, jumping hosts horizontally and then melding with the vertically transmitted information (a pattern that only rarely occurs in biological mutualism). In the shrunken world created by Western exploration, exploitation, and communication, the latter epidemiological pattern of course has come to the fore, but it is the former cladistic pattern that was perhaps dominant before the age of the empires.

In the rest of this chapter, I want to illustrate some of these essential properties of culture by drawing on just one culture that better illustrates the nature of cultural entities in human prehistory than our own gigantic, rambling, fast-changing conglomerate. Although none of the following observations are startingly original, I hope they will serve as vivid reminders of points that we often forget. Our explanatory models for coevolution can improve only if we have the explicandum, the nature of traditional cultures, constantly in mind.

An Example: The Culture of Rossel Island, Papua New Guinea

Rossel Island lies about 500 km off the coast of New Guinea, the easternmost island in the Louiseade Archipelago. It is a small volcanic or "high" island, 34 km long by 14 km wide, with a central mountain range 850 m high that is clad in rain forest. Despite fluctuations in sea level, it was not connected to other islands during the Pleistocene, and consequently has a relatively limited fauna and flora compared with the mainland (see Mayr and Diamond 2001 on the Melanesian island world). It is separated from the nearest other islands by difficult waters full of reefs, for which reason there was little contact with Western shipping until the early 1900s, with a resident Catholic mission only set up in 1953. It has a small human population; in 1920 there were 1450 Rossel islanders, while the current population stands at 3884. Such a small population, relatively cut off from migration, is likely to exhibit both founder effects and genetic drift in its population biology.[8] Small, isolated populations of this sort are vulnerable to natural disasters of various kinds, including diseases, cyclones, and drought—not to mention potential conquest from more numerous neighbors. Physical anthropologists have shown that the inhabitants of Rossel Island are shorter (at c. 155 cm for males) and darker and less brachycephalic than neighboring populations (Armstrong 1928). A recent study of genetic markers shows that they are not close kin to most of their neighbors, but rather to highland mainland New Guinea populations (M. Kayser, personal communication). Biologically then, the Rossel Island population is distinct from the populations of most Oceanic islands, which derive more directly from an Asian stock associated with the spread of Austronesian languages throughout the Pacific from about 4000 BP.

Culturally, too, Rossel is distinct. The inhabitants (who I will sometimes refer to as Rossels) speak a language they call Yélî Dnye, which is not clearly related to any other existing language, and certainly not to the surrounding Austronesian languages. It has many unusual complex properties to which I will return later. The islanders are culturally distinctive in many other respects too. Compared with their Austronesian neighbors, they have a quite different musical system, distinctive canoes and houses, their own indigenous shell money system, and a dual-descent kinship system that contrasts with the matrilineal descent systems on neighboring islands. Neighboring peoples point to the absence of traditional pottery, drums, carving, tattoos or other forms of visual art, and to the peculiar and apparently unlearnable language, and they regard Rossel culture as a thing apart. The presumption must be that Rossel Island culture is an ancient cultural continuity, a remnant of the pre-Austronesian offshore cultures that we know from radiocarbon dates to have inhabited the islands of near Oceania more than 30,000 years ago (Kirch 1997; Spriggs 1997). It is not of course a frozen cultural relic. Closer examination shows many borrowed cultural traits, and the culture has no doubt developed in its own distinctive way over the intervening millennia (and most recently under colonial impact), but it seems indubitable that it descends from cultures that were in the area before the Lapita peoples speaking Austronesian languages passed through 4000 years ago. From that pre-Austronesian time the culture almost certainly inherits taro and sago cultivation, nut-processing technologies (see later discussion), patrilineal inheritance of land, many properties of the language, and quite plausibly, cultural patterns of house construction, a musical system, and so forth.

With that preamble, I would now like to exemplify some of the key properties of culture listed earlier, in order to put some flesh on those bare bones. We will take the features in turn.

System Complexity Without Any Single Designer

I propose the following hypothesis: Cultures will tend to become more complex over time, up to the limits of the transmission process (where transmission is constrained by what is individually learnable on the one hand, and by social and cultural constraints, such as the degree of division of labor, or whether the society is literate, on the other). A fundamental reason for this is that in a culture, the lack of fidelity in reproducing ideas and practices is much less likely to be "fatal" to the continuation of the practice than the high probability that a mutation will be fatal to an organism. Consequently, nondeleterious innovations and variations can accumulate, even though they may require constant adjustments of the system in which they occur (cf. how sound changes in language can require distinctions being lost in one place to be remade in another, as in the English great vowel shift). Between innovations themselves and the adjustments of the system that are then required, complexity can accumulate. Such complexity will be eroded, not just by transmission error (for example, by children's simplifications while learning), but also more systematically by contact with other cultures and creolization. Contact, trade, and cross-cultural communication are levelers of cultural distinctiveness, as exemplified by the current spread of Western practices across the world.

Given the inaccessibility of Rossel Island, its cultural complexity has extended to somewhere near the limit for a society without literacy or a significant division of labor. As a start, take the language. It has the largest phoneme inventory (ninety distinct segments) in the Pacific, and many sounds (such as doubly articulated labial coronal stops) that are either unique or rare in the languages of the world. Among the fifty-six consonants are many multiply articulated segments; e.g., /ṭpn̩m/ is a single segment made by simultaneously putting the tongue behind the alveolar ridge, trilling the lips, and snorting air through the nose. Such multiple articulations are a formidable barrier to the learner since different emphasis on one or the other articulation can give a quite different auditory flavor, sounding more like a /t/ or a /p/ or an /n/. Once the learner is past the sound hurdle, he or she faces another formidable obstacle. The language has an extremely complex system of verb inflection (with thousands of distinct inflectional forms). For example, the properties of the subject (singular, dual, plural, first, second, or third person) of the verb is marked before it, but with a single syllable that also encodes tense, aspect, and mood; altogether there are 144 combinations of these. In addition, substitute forms are used where the subject has been mentioned before, is close or visible, is in motion, or where the sentence is counterfactual or negative, thus providing well over a thousand possibilities. Meanwhile, after the verb, another particle marks a lot of the same information, together with the person and number of the object. To reduce the combinatorial explosion of distinctions, all nine possible subject combinations (e.g., second person dual) are grouped into two categories, "first person or singular" versus. "second or third person dual or plural," called monofocal and polyfocal, respectively, in the Papuan linguistics literature).

This particular and unusual kind of grouping happens to be found also in the Gorokan languages on the mainland, indicating a possible distant relationship to peoples more than 1300 km away. The point is that here is a complex but logical and consistent way of cross-referencing subject and object (together with tense, aspect, and so forth) on the verb; the subject is categorized twice using different categories, once in front and once behind the verb. It is an intricate piece of clockwork designed by "the blind watchmaker"; that is, by eons of use by generations of individuals whose tiny unseen slips and innovations have been sculpted into a functional system by the selective forces of learners and users of the language.

Yet another barrier to the learner is that most verbs supplete (varying in root like English *go* versus *went*) in many grammatical contexts. The overall result is that the language is at the boundaries of learnability. Hardly any mature individuals (such as non-native spouses) who have immigrated into the island community ever learn to speak the language, and children of expatriate Rossels do not fully acquire it from their parents alone.

This kind of intricate complexity is familiar to scholars who work on the languages of small, relatively isolated indigenous communities, whether in the Americas, Australia, or New Guinea. In the case of Rossel language, we know it has been cut off in a sea of Austronesian languages for more than 4000 years, and has been left as it were to develop on its own, in the inevitable direction (I am suggesting) of complexity. In contrast, languages that function as a lingua franca across borders and boundaries, such as English, Indonesian, or Spanish, cannot sustain such complexities (at least in their interethnic uses), tending toward the simplification found in creoles. They are leveled by the need for commonality, and in many cases by the fact that they are learned as second languages for limited functions.[9]

Cultural complexity on Rossel can be found in many other practices. The kinship system is effectively a dual-descent system in which individuals trace their membership in both matrilineages and land-bearing patrilines. The kin terminology, with more than forty terms, is classificatory; that is, all terms apply, not just to a single individual (such as English "mother"), but to a large class of individuals (such as English "cousin"). There are three distinct ways to decide who falls into a class. You can reason genealogically; e.g., my mother's mother's brother is a *mbwó*, just like my brother. You can reason by clan membership; any male of my matriclan of my generation or an even generation counting from mine (e.g., two generations up or two down) is a *mbwó*. Or you can reason relationally; if my *pye* (someone in the class of my mother) calls someone *kênê* ("mother's brother"), then I can call him *mbwó*. Now here is an interesting case of design without a designer. Here is a calculus of relationships that will unerringly assign individuals to the same class by different rules, while keeping the whole system objectively coherent, so that if you call someone *kênê*, then that is consistent with my calling him *mbwó*, given that you are my *pye*. Incidentally, all these rules change if you are a female calculating the system from your point of view, but the end product is consistent with the system used by your brother.

Working the system requires incredible genealogical knowledge. Mature Rossels know their own descent lines up to ten generations deep, and in many cases more or less have a command of the entire genealogical relations between any two individuals on the island. Knowing these relationships is essential to the proper use of kin terms, the assertion of rights, plans for alliance and marriage, and correct deportment and the appropriate use of taboo language (alternative words for items such as clothes and body parts, which are used in the presence of in-laws).

Another cultural institution is an indigenous shell money system, famous in anthropology as the most complex indigenous system recorded (Armstrong 1928; Liep 1983a), with about twenty named denominations of shell coins in two parallel series. The purchase of a pig may involve up to 1500 coins of one series and 800 of the other, a quantity that no man has in his pocket (or rather basket, as appropriate on Rossel Island), and which must be assembled by an elaborate system of loans and securities requiring hundreds of transactions following specific cultural rules. High-value coins can be borrowed only by presenting the next lower denomination as security, the loan of which in turn will require its own security and so on down the line. Purchases thus mobilize extensive networks of kin and business associates, in a manner reminiscent of the joint funding of some enormous engineering project like the Channel Tunnel. Again, we have a system of great intricacy that has no doubt evolved culturally over thousands of years—a system whose design the Rossels attribute to the gods.

It seems that many Rossel cultural institutions have developed a complexity that approaches the limit for cultural transmission across generations in a society without literacy or even any significant division of labor. Every adult knows the same essentials, with expertise confined to those of relatively advanced age (in a society with an average life expectancy of about 45 years). What is complex is not of course merely the constituent ideas (the memes if one will), but their articulation into a functional whole.

Multiplicity and Variation

It is a commonplace that cultures vary. Yet anthropology textbooks are replete with generalizations about institutions across unrelated cultures, be it kinship, witchcraft, or warfare. Although few anthropologists agree about the details of such generalizations, the overall picture is nevertheless clearly one of variation within constraints. Let us concentrate first on the variation, the uniqueness of particular cultural institutions; however, it is important to not lose sight of some of the remarkable similarities.

So far, we have seen intricate cultural systems on Rossel Island, which have evolved to fulfill precise design requirements without any mastermind behind it all. How have they arisen? By cultural evolution of course; that is, by selection of variants over generations until the systems have come to fulfill ever more complex functions. That presupposes variation. Can significant cultural variation exist in a population of only 4000? Yes it can. Take the language. There are two main dialects, an eastern and a western, differing in syntax, morphology, and lexicon. Now take the eastern dialect. It is separated into a northern and a southern variety, divided by a mountain range. Now take the southern variety. It has an eastern and a western subvariety. Now take the western subvariety. Small differences can be found among most villages. There is variation all the way down. Similar variation can be found in most other cultural practices. A Rossel islander looking at a sago-processing device (a chute with a funnel and a strainer, all made from local bush materials) can tell where the maker came from, for there are "dialects" of sago-processing devices. The same is true for a host of other cultural features, from baskets to song styles.

There are of course exogenous sources of variation also. After a storm, a canoe from another island sometimes washes up on the beach, spurring experimentation with canoe design. Visiting traders come seeking exotic marine produce for sale as aphrodisiacs in Asia, supplying diving goggles which spur new fishing practices. Young men return from the mainland with messianic religious ideas or the concept of home brew. A woman from another island who has married in introduces a new kind of basket. And so on (see Liep 1983b for the history of colonial influences).

Small indigenous communities, despite the "traditional" epithet, in some ways exhibit a wider range of variation in cultural practice than that allowed by the solutions to practical problems found in industrialized societies, where language is standardized, clothes come off the rack, bread comes from a large-scale bakery, or entertainment arrives via an electronic tube. Since the existing strands of variation are the stuff on which cultural selection works, traditional societies contain the resources for rapid change if it is required.

Given cultural evolution, why can we discern clear commonalities and resemblances across cultures, as in kinship, religion, and political systems (admitting of course that there are many distinct types)? One source of convergence is what we might call the cognitive bottleneck. Cultural elements have to be learnable, memorizable, and computable on a reasonable time scale. They also have to conform to our motivational propensities.[10] There are also sociological, economic, and ecological constraints. Many conceivable cultural systems (pure communism, for example) just wouldn't work.

Another source is cultural borrowing (see the section on horizontal transmission). And there is always the possibility of inheritance from a common ancestor, as discussed next.

Vertical Transmission and Cladistic Character

I have already outlined the way in which the culture and language of Rossel Island is distinct from the Austronesian languages and cultures on the nearest islands. The relationship between those Austronesian languages is now well understood (Lynch et al., 2002), and a family tree can be reconstructed from the papuan tip cluster (the languages surrounding Rossel Island) up to the higher western Oceanic grouping, then up to proto-Oceanic and all the way back to an Asian proto-Austronesian. In some cases one cannot be sure whether one is dealing with sister languages or languages that have converged by borrowing at a later date, but generally the cladistic pattern is quite clear. In contrast, in the offshore islands to the east of Papua New Guinea, including the Bismarks, Bougainville, and the Solomons, there are about thirty non-Austronesian languages, including Rossel's Yélî Dnye, whose relationships to one another, if any, are much less clear (these are so-called Papuan languages, a term hat might misleadingly suggest a language family, but in fact only means they are non-Austronesian). The very fact that we cannot reconstruct any certain relationships between them suggests a much greater time depth for this wave of human settlement than the Austronesian spread of c.3000–4000 years ago. As mentioned, we have radiocarbon dates from the Bismarks that go back to 35,000 BP, and from Buka (then part of a Greater Solomons landmass) that go back nearly to 30,000 (Kirch 1997; Spriggs 1997), so we must assume that most of the islands of near Oceania were settled in the Pleistocene. Recent evidence suggests that by 20,000 BP these peoples were cultivating taro and various nut species (Spriggs 1997: 38), exploiting an introduced or semidomesticated arboreal marsupial (the gray cuscus, *Phalanger orientalis*), and trading obsidian for edged tools (Kirch 1997:35). Current speakers of offshore Papuan languages may be descended from these early colonists, or from later waves of immigration in the early Holocene. But the high estimation of taro, nut, and cuscus as foods survives in the modern Rossel value system, suggesting some ancient cultural continuities.

I have mentioned that recent genetics ties Rossel islanders to the eastern highland populations of New Guinea over a thousand kilometers away, rather than to the neighboring islanders who speak Austronesian languages.[11] I have also mentioned that there is the occasional linguistic feature (e.g., the classification of first and singular persons in monofocal verbal inflections) that suggests just such a tie, in this case to the Gorokan languages. Perhaps in the long run we will be able to establish connections both to the main island and to other offshore Papuan languages (see Terrill et al. 2002). In the meantime what is clear is that this small population has retained many cultural features over deep time, developing them to an extreme of complexity, through vertical transmission over the generations via implicit learning and explicit instruction. Children accompany adults on almost every kind of venture or expedition, and they thus have an extensive understanding of most aspects of the cultural system before they reach puberty.

Horizontal Transmission and Cultural Diffusion

Isolated though it is, over millennia Rossel Island has no doubt had plenty of visitors, not always intentional, because the reefs have sunk many a ship. Even without visitors, wandering Rossels have returned with alien ideas, or ideas embodied in cultural objects have washed up on the beaches. The cultural repertoire includes many features similar or identical to those on neighboring islands. Much of the agricultural system, some of the house and canoe styles, even the matrilineal clan system with bird totems is shared throughout the Massim area. Today the inhabitants have been missionized for half a century, sing English hymns in church, wear Western clothes (secondhand from Australia), and use metal tools and nylon fishing lines.

To some limited extent we can reconstruct the contacts with the outside world before Western ships first showed on the horizon in the eighteenth century. In precolonial days, legends recount trade with the neighboring Sudest Island, controlled by a few a "big" men or political leaders. From Sudest came clay pots, stone ceremonial axes, and the plumes of birds of paradise; in return Rossel sent precious shell necklaces for use in the Kula trade (Liep 1983b). Further back in mythical time, Sudest is held to be the source of various important cultural items, including the dog, the yam, and the sailing canoe. Although such myths cannot be taken as history, they do consistently paint a picture of Sudest as the source of many cultivars and elements of technology. Since there has been no archaeological investigation on Rossel, we cannot date these imports directly.

However, the language gives us important clues. Although the lexicon shows very few loans from Austronesian languages, those words that have been borrowed tell a fascinating story. Take the words for numbers. The language of Rossal Island has a full-scale decimal system that is entirely regular in construction (100 is denoted by "the tenth ten," 1000 by "the tenth, tenth ten," and so forth). The words clearly show their Austronesian origin. For example, Rossel *peeti* or *paati* ("four") is clearly derived from proto-Oceanic **pati*, Rossel *limi* ("five") from proto-Oceanic **limá*, and so forth. Because the language of Rossel Island has a huge phoneme inventory, loans are apparently not corrupted but are represented faithfully. Note that we also have Rossel *waali* ("eight") for proto-Oceanic **walu*. The interest of this is that in most of the surrounding Austronesian languages, "eight" is represented in a different way—as "five plus three"—because the Papuan tip cluster subbranch of Oceanic to which these languages belong innovated such a "five plus" system about 3000 years ago (Lynch et al. 2002). Thus Rossel language borrowed the number "eight" before that subbranch spread. There must have been earlier Oceanic peoples who were in contact with the Rossel people. A few aspects of proto-Oceanic are better represented in Rossel language than in the local Oceanic languages that are now neighbors today!

If we look at the other Oceanic loans in Rossel language, we find words for aspects of material culture, such as "pot," "bottle," "lid," "clay pot," "woven coconut mat," "grass skirt," "armband," "shell necklace," and so forth—items that we may assume were borrowed with their names.[12] The lime used to get the maximal "kick" out of chewing betel nuts was also clearly a cultural borrowing; the words for the white branching coral from which it is made and the lime pot it is kept in are loans. Other loans reveal the borrowing of seafaring knowledge and equipment; the words for "sail," "wind," "westerly wind," "fish poison," etc. are also Oceanic loans. If we believe this evidence, Rossel islanders before, say, 3000 years ago lacked cooking and storage vessels, woven mats, and various kinds of clothing and adornment. They may also have lacked the main stimulant of today, the betel nut,[13] and they seem not to have had sailing canoes, and perhaps had very modest maritime technology.

What use did the inhabitants of Rossel Island have for a counting system that runs smoothly into the thousands (mainland Papuan languages often have a body-based counting system that terminates at, say, 31, on the navel)? To this day it only has one predominant use: counting the shell money mentioned earlier. Was this money system also a cultural borrowing? Shell beads that date back to 8000 BP have been found in archaeological sites in the Bismarks (Spriggs 1997: 59), and shell armbands occur in pre-Austronesian sites in the Solomons (Kirch 1997: 41), but a preoccupation with shell valuables is a distinctive feature of the Lapita culture (starting c. 4000 BP), which is presumed to belong to the first Austronesian-speaking peoples of Oceania (Kirch 1997: 236–238). It seems likely then that the shell-money system, now a distinctively Rossel cultural trait, was at least elaborated through contact with Oceanic peoples.

In trying to trace the diffusion of cultural traits, one has to take into account the possibility of independent invention. Take the polished stone axe, not chipped but ground down in the "neolithic" manner. Cultures all over the world, from South America to Australia, seem to have developed the polished axe, long after the dispersal of humans, and when they had long been out of contact with one another. [In fact, the earliest known examples probably come from Pleistocene New Guinea (Spriggs 1997: 59), and the same design was in use on Rossel Island until about 1900.] Yet the axes look almost identical, whether they are made in Mexico or Queensland. Somehow the functional requirements, given human cognitive and anatomical skills, converge on a single optimal design in Dennett's (1995) design space. Such examples of con-vergent solutions are numerous: the ladder, the basket, rectangular houses with pent roofs, even the idea of domestication of local plants. In language, for example, there are restricted types of design in many parameters. For example, case systems are either nominative-accusative (where subjects share one case and objects another, such as *he* versus *him*) or ergative-absolutive (where the object of a transitive verb is the same case as the subject of an intransitive verb). Cultural convergence is just like parallel evolution in biology, where striking parallels in design arise independently. In the case of culture, these tell us something important both about constraints on cultural design, which is perhaps quite largely cognitive in nature, and the tendency toward optimal design, and thus about the ceaseless selective pressures in the evolution of culture.

Distinguishing diffusion from parallel invention can be difficult. Perhaps given time, Rossel islanders would have invented the ladder themselves, but it is just as likely that they saw a specimen before they got around to inventing one independently. Take Rossel language. It has an ergative case system, but so do perhaps a quarter of the world's languages, scattered from the Caucasus to Middle America. However, this case system in Rossel language also makes a difference to the syntax, so the language can be said to have an ergative syntax. That is much rarer; perhaps less than 2 percent of languages with an ergative case have an ergative syntax, and there are only a few known pockets around the world—one in Mesoamerica and one in Queensland, just 500 miles away across the ocean from Rossel Island. Is this a trace of ancient diffusion, or even ancient inheritance from a Sahul ancestor (Sahul is the name of the archaic continent that included New Guinea and Australia until 9000 years ago)? Who knows?

The general point here is that even the most traditional of cultures are porous to new ideas, especially, as we have seen in the Rossel Island case, ideas that are in a broad sense technological, including what Goody (1977) has called "the technology of the mind" (exemplified here by the decimal system). It is this that gives cultures the ability to change rapidly and to adapt to the greatest danger that faces any human group, namely, other human groups.

The Cumulative Nature of Cultural Transmission

Nothing perhaps tells us more about the virtues of Lamarckian inheritance than the procedures cultures employ for exploiting the environment. Consider the wild nuts collected and eaten on Rossel Island. Some of these show up in archaeological deposits on the mainland well before the Austronesians arrived (Kirch 1997: 39–40), so the practice of exploitation is ancient. Now consider that many of these are deadly poisonous if they are eaten before processing. On Rossel Island, the *kwee* nut, for example, is the fruit of a large vine, yielding large 5-cm nuts in giant pods. However, if it is eaten raw, the *kwee* nut is fatal. It has to be deshelled, the kernels roasted until they are soft, then pounded and placed in an open-weave basket in a fast-flowing river for 5 days. Two boys who in 1997 stole some nuts from the river after only 3 days of leaching almost died, frothing at the mouth; they were saved by the administration of the seeds of another plant (the antidote *polo*), which made them vomit. Rossel islanders exploit half a dozen species of forest nut that need processing in a similar manner, but some need longer roasting or leaching for only 2 days, and so forth.

Experimentation with foodstuffs in the tropics is a hazardous enterprise. Tropical plants have evolved all sorts of chemical defenses against the huge and bountiful insects that would otherwise chew them up. Even walking through the bush requires care because the leaves and sap of many trees are caustic. Some of these poisons also have their cultural uses. For example, three plants are used to poison fish—the roots of one, the leaves of another, and the fruits of a third. These three plants are carefully planted in the bush and nurtured, ready for use. They are crushed and are sufficiently powerful that a basketful will poison an entire river, sending its fish gasping to the surface, so that a whole village can collect the bonanza. In addition, many such plants have medicinal properties if they are used in carefully controlled proportions known to specialist medicine men.

Rossel Island is surrounded by magnificent reefs inhabited by more than 1000 species of fish—the richest marine life in the world. Indigenous knowledge about fish is highly developed. There are hundreds of single-word names for species and genera of fish, with recognition that juveniles of the same species can look very different from adults. Some fish are known to be poisonous. A species of pufferfish (*mt:enge*, an Arothron species) is carefully avoided as deadly poisonous at one end of the island, but at the other end is eaten after the poison sac is extracted. More intriguing perhaps is the treatment of big game fish like barracuda, which can accumulate ciguatera poisoning (by ingesting reef fish that have themselves ingested poisonous *Gambierdiscus* or *Ostreopsis* algae). The symptoms include nausea, loss of sensation, and occasionally death. The islanders treat such fish with caution. Large specimens are rejected; smaller ones are carefully gutted, skinned, and boned and then tested. If the flesh is very firm, they are rejected; if not, the raw flesh is put on the lips and if any tingling or pain is felt, the fish is rejected. Only fish processed and tested in this way are eaten, and cases of poisoning are then rare.

These examples make obvious the enormous and crucial value of accumulated, traditional knowledge. No individual could hope to test all these essential sources of nutrition and survive. Although bush foods are not crucial if root crops are available, they are essential in times of drought or between harvests. Arriving on such an island, Robinson Crusoe would either starve or rapidly poison himself. Of course such accumulated knowledge is vital in many other cases as well: being able to "read" the weather, the tides, and the winds is essential to safe canoeing or fishing. Discerning the best soils is essential before starting the labor-intensive task of felling the rain forest if next year's crops are to be successful. More than 200 kinds of bush plants and trees are exploited for building houses, making ropes, constructing canoes, making glues, or weaving baskets. Every individual may experiment a little bit, but it is done within the guidelines and with the safety net of millennia of accumulated wisdom about the local environment. We tend to forget that many domesticated species, such as almonds, lima beans, cabbage, potatoes, tomatoes, rhubarb, watermelons, and eggplants derive from poisonous or intensely bitter wild ancestors (Diamond 1998: 118). Some major domesticates remain poisonous until they are cooked (potatoes if at all green or sprouted) or specially treated (manioc). Others require major processing to extract the nutritive content, such as sago, where the starch is leached out of pulverized palm trunk. Such cultural knowledge obviously unlocks natural resources and enhances biological fitness, building a much larger, healthier population than would otherwise be possible in an environment like Rossel Island.

Cultures Include Functionless or even Deleterious Elements

The annals of anthropology are full of records of apparently irrational behavior, magic rites, rituals, unnecessary violence and warfare (and plenty of attempts to show that, given the native views, such madness maybe makes sense after all). It would be fairer to focus on some of our own idiocies, but I will stick to plan and give one example from Rossel Island. The islanders were cannibals up until the 1920s at least. To this day, they believe that no one dies naturally, but is the victim of sorcery. If an important man or chief died in the past, the sorcerer was identified by divination, and made to produce a victim to be eaten at the mortuary feast for the chief, the relatives receiving high-value shell money in compensation (Armstrong 1928: 103–114). There were also other occasions for cannibalism, including the consumption of shipwrecked aliens and the eating of persons who had defiled sacred places or otherwise caused trouble. The practice of cannibalism is now dead (in fact it seems to have been gladly dropped in response to colonial pressure), but the underlying ideas about sorcery remain.

I have yet to find a Rossel islander who does not believe in sorcery. It is a universal belief, and every death proves the point. Even murders are attributed to sorcery; the murderers were either themselves sorcerers who incapacitated the defenses of their victims, or coerced others to do the deed by threatening sorcery. There is a shrine to a spirit who inflicts countersorcery. The bereaved may go there and purchase revenge from the keeper of the shrine; the spirit then inflicts the sorcerer with cancer or another wasting disease like tuberculosis. Anyone dying of a wasting disease in a village where there has been an earlier death can be assumed to be the sorcerer who caused the earlier death; again, the one death proves the truth about the other. Specialist diviners and orators skilled in veiled accusations make a career out of the system.

This self-justifying system of beliefs casts a pall over every otherwise happy circumstance on the island. People go to intracommunal feasts and dances with both happy anticipation and watchful anxiety, for communion with strangers gives a sorcerer opportunities. In the same way, venturing into the hinterland alone or at dusk is full of dangers. Pregnancy, birth, and childhood are also times of joy and special danger. Visitors to other villages hasten to the relative safety of their kin or clansmen, taking special care with food. Suitors and those on "banking" expeditions for shell money travel full of trepidation and precautions. In short, social intercourse on the entire island is constrained and limited by fear of sorcery. No wonder dialect differences can flourish despite geographic proximity.

I take this as just one example of a deleterious cultural practice that may in the past have had biological consequences, since eating human flesh can pass on fatal diseases (see Durham 1991: 393–414 for an overview of the effects among the Fore of highland New Guinea). Existing beliefs about sorcery are also deleterious, since they restrict the circle of marriage, lessen trade and economic exchange in times of dearth, and undermine joint ventures and political alliances. Since a cultural system is a web of beliefs, some of which (like the poisonous nature of certain foodstuffs) are better not directly tested, once a deleterious practice becomes embedded in tradition, it is hard to eradicate. In traditional societies, the very mechanisms that guarantee a probability of vertical transmission will protect useless or even harmful beliefs from rapid elimination.

Culture, Group Selection, and the Feedback to Biology

As discussed earlier, group selection is a controversial mechanism in evolutionary theory. However, it seems odd to doubt the efficacy of culture in political decisions that affect the life chances of entire groups. In March 1997, cyclone Justin hit Rossel Island and circled over the island for 4 days, dumping more than 3 m of rain and whipping the island from every side with winds of up to 250 km an hour. At the time, I was on a neighboring island on my way to Rossel, and I was thus able to observe the immediate aftermath at first hand. Sailing toward the island a few days later, the sea was glassy smooth but full of logs, dead fish and crocodiles, and other debris. When the island came into sight, the tropical forest, normally intensely green, was entirely brown, denuded of leaves and burnt by salt spray right up to the central mountain peaks. Landslides triggered by the colossal rainfall left huge gashes on the mountainsides. The lagoon was brown with sediment and the reefs were covered in sand. Indeed, the whole local topography had changed, with new islands on the reef, rivers having changed their courses; the rain forest was flattened, and stretches of mangrove forest were washed away. Remarkably, there had been little direct loss of human life, despite the fall of many large trees, landslides, and the collapse of most of the houses, with even parts of coastal villages swept out to sea by huge waves that came over the barrier reef.

In fact it turns out that Rossel Island culture is adapted to cyclones, which are experienced as direct hits once or twice in a decade. Every village has a cyclone house, a barrel-shaped long-house of sturdy construction built directly on the ground, unlike everyday houses, which are raised on stilts. These are not a recent innovation—they are mentioned in 1849 as an unusual feature observed from offshore by Thomas Huxley and crew, who were unable to land on the island (see Armstrong 1928: 14–15); indeed Rossel islanders hold that the cyclone shelters were designed by their principal deity, N*gwonoch:a*. They are only 2 m high, anchored to the ground by sturdy piles, roofed with thick sago leaf, and equipped with benches raised off the floor in case of flooding. They are jointly maintained as a community resource.

There was an immediate and decisive community response to the disaster of cyclone Justin. Virtually all the normal food crops had disappeared in floods and landslides, leaving scarcely any seed tubers. The islanders know well that in such circumstances they rely crucially on sago. Sago is the starch leached out of the pith of *Metroxylon* palms. It is processed by pouring water against the pith held in a sieve or net; the starchy water is evaporated to yield a flour, all utilizing another elaborate indigenous technology. The cyclone knocked down all the mature sago trees, which are privately owned. Such felled logs must be processed in a few weeks before they mildew and become infested by pests. The community response was immediate; private ownership was waived and communal teams were set to extracting the sago starch, which when dry can be kept for about 4 months. This was to be the main means for survival.

There were other essential community responses. Drinking-water sources were fouled by salt spray, dead animals, and fish; the best sources were selected and communally cleared of debris and cleaned out. Community labor was used to clear villages and repair houses where possible. Thatch was in short supply, and communal labor was used to retrieve the sago leaves that lay on the ground and to collect a wild substitute from a short palm in the bush, although individuals thatched their own houses. Those whose houses were unrepairable lodged with their neighbors.

The extent of the disaster now became clear. Covered in sand, the reefs no longer sustained a rich marine life, and offshore fish and shellfish had effectively disappeared; indeed marine life only slowly recovered over the following year. Some fish, shellfish, and crabs could be found in the mangroves, but protein was now scarce because most of the pigs had perished in the storm or gone wild. Coconuts, an important source of fat, would not be available for a full year either. Sago palms with their essential thatch and useful starch, would not be available for 3 years. What about bush foods? The rivers had changed their courses, and the usual supply of prawns, eels, and fish was gone. The dense rain forest, as mentioned, was stripped of leaves by the wind; those surviving turned brown and dropped from salt burn coursed by windblown spray. Without the leaves, the forest floor dried up and the fallen leaves rotted like a giant compost heap. No rain fell for weeks. Those few seed tubers that had survived were treated as communally owned and carefully planted in the soil. Unfortunately there was none of the usual nurturing shade and moisture, and the soil was poisoned by salt. Then just as the rains returned, there was a plague of caterpillars that ate every shoot that sprouted. The caterpillars would normally have been happily occupied eating the leaves of the forest, but now the entire insect population of the island had only the seed tubers to feast on! (Many of the insect species that now became apparent seemed entirely new to the islanders, who imagined they had been transported to the island by the cyclone.) Community teams daily set out to crush the insects underfoot. The caterpillar plague peaked about 3 months after the cyclone, by which time the forest trees were tentatively putting forth leaves again, but it would be a year before bush nuts were available. Then the surviving wild pigs descended on the gardens in search of the precious tubers, making only the gardens closest to villages defensible. Community hunts with packs of wild dogs strove to hold the pigs back in the mountains.

It is in this context that the wisdom of the intense communal response immediately after the cyclone becomes clear. Without the requisitioning of the sago, there would have been large-scale famine, and without communal efforts to save the seed tubers, there would have been no following generation of garden produce. The cleaning of the water holes prevented disease, which might easily have swept through the communities. In the event, the population was malnourished for 6 months, but by November it was on the road to slow recovery. In the interim, a second calamity—drought, insect plagues, or an epidemic—could have had disastrous effects. Captain Cook recorded that the 1775 cyclone that hit Pingelap in Micronesia reduced the population from 1600 persons to a mere 30 by 1777 (cited in Cavalli-Sforza et al. 1994: 352). The fate of the Rossel island population at this critical time hung on a thread. What saved it was coordinated communal effort, which was made possible by an effective political system built on a local variant of the "big man" system (persons who acquire authority through effective use of exchange and the distribution of resources). Here the indirect benefits of the shell-money system can be seen; " big men" acquire their authority especially through control of shell money, which is required for marriage, death, and major traditional undertakings. Young men in search of wives thus affiliate with these elders, who can direct their activities and so wield real power.

I have recounted this episode in detail because I think it makes a clear case for the role of the group, as an effective joint "superorganism," in the biological survival of the population. The effectiveness of the group depends on its role as a culture-bearing entity with transmitted traditions about how to prepare for cyclones with cyclone houses, how to deal with the challenge of storing enough food (requisitioned sago) to stave off starvation during the immediate period of shortage, and how to overcome private interests (in sago trees or seed tubers) through the exercise of political power in the interest of group survival. A group without these mechanisms is not likely to muddle through such a disaster. However, these mechanisms also depend on biological underpinnings: the ability to inhibit private interest, to pull together in times of stress, and also more generally the cognitive abilities that make long-term communal planning possible.

Still, many will feel that the feedback from culture to biology is a thing that happened in deep prehistory (hence the popularity of the evolutionary psychology manifesto of "space age man with a stone age mind"). Nevertheless, microevolution, with feedback loops, is discernible in the course of just a handful of generations. Take the case of deaf communities, which usually support a manual sign language. Recent research has established that sign languages of this kind are full-fledged languages, with all the expressive power, syntactic flexibility, and lexical richness of any natural language. In Europe and America these languages have become relatively standardized in institutions for the deaf, but in other lands they have evolved more haphazardly wherever a critical mass of deaf people have come together. A number of village sign languages have been reported where a persistent strand of hereditary deafness has consistently generated a number of generations of deaf individuals. On Rossel Island there is such a strand of hereditary deafness, where a number of families have three generations or more of deaf individuals. Not only these individuals, but also the hearing members of the family and indeed all members of the villages where they live have developed a sign language for effective communication. This system, like most of the other reported village sign systems, remains to be scientifically researched. My initial investigations of one such family with three deaf adult children show that the sign system is capable of conveying quite abstract messages; for example, about events in the future, things witnessed in the past, or hopes and desires in the present. By virtue of the developed sign system, the deaf members of this family are fully integrated members of the village community. Two of them have married in the traditional way, involving complex exchanges of shell money between kin, and have children, some of whom are deaf, so the sign system will have a future utility, and is in effect a strand of cultural tradition in the making.

Now consider for a moment the interaction between the biological and cultural elements here. A strand of hereditary deafness prompts a cultural development, a systematic system of manual signs. In turn that systematic sign language renders the deaf individuals fully functional members of society. Their extended kin therefore invest them in the normal way, providing the shell money necessary for marriage. This legitimates biological reproduction, which in turn gives continuity to the genes that generate the strand of deafness. This is a microsystem of gene–culture coevolution over a handful of generations whose properties can be directly studied (see Aoki and Feldman 1991).

Conclusion

In the first part of this chapter, I outlined the kind of general considerations involved in trying to understand the evolutionary background of culture. There I hope to have established that twin-track theories of gene–culture coevolution are perfectly plausible, and moreover do not depend on the meme, that is, on a culture particle, as it were, mimicking the gene. Just as the soft Darwinists imagine natural selection operating on many levels, selection of cultural forms can apply at any level, from a minor design feature (such as the curve of a canoe prow) to a whole system of interrelated parts (such as a kinship system or a language). In a phylogenetic perspective, the crucial locus of gene–culture coevolution is cognition, because without the cognitive foundations for culture (abilities to "read" other minds, to inhibit actions, to learn any number of cultural variants in some domain), the accumulative character of human culture could not have gotten off the ground. In an intraspecific perspective, cultural variety allows geographic radiation into specialized niches, where microevolution in response to particular conditions (extreme cold, malaria, lack of salt) can arise. In a comparative cultural perspective, once cumulative aspects of culture start to pervade the environment, they in turn act as a selecting environment acting directly on the genotype (as illustrated by, e.g., the systematic association of dairying cultures and the genes for lactose absorption). In an intracultural perspective, genetic diversity within a cultural group can lead to microevolution of culture, as illustrated by sign languages of the deaf.

In the second half of the chapter, I have given a set of reminders about some of the crucial properties of culture that need to be modeled by any encompassing theory of cultural evolution. The target for explanation, a particular culture, has particular properties. It is not a heap of traits, but a system of systems of amazing complexity; there is variation all the way down to the individual. Some properties of individual cultures have extraordinary robustness to change; others are subject to rapid change, borrowing, and cultural exchange. Cultures accumulate successful trials of myriads of adaptive experiments (as in the example of processing of poisonous foods), but they also accumulate nonadaptive or even deleterious features. Aspects of culture can confer adaptive advantage on entire groups, and special microadaptations of culture to genetic impairments can evolve rapidly. I illustrated all this with examples from a relatively isolated society of simple technology, the kind of society that predominated through human prehistory.

I would like now to try and draw these two strands closer together. The parallel between genetic and cultural evolution sometimes seems forced. Gould (1991: 63) went so far as to say that "I am convinced that comparisons between biological evolution and human cultural or technological change have done vastly more harm than good." However, there are lessons to be learned from the exercise. One is that conservatism, i.e., resistance to change, is essential to the growth of complexity. In genetic evolution, conservatism is built in through the fidelity of reproduction; mutations are the exception, and most of them are extinguished. If the meme was the full story for culture, cultural change as a whole would be as rapid as fashions in clothes or pop songs, and in that case "ratchet culture" could not exist. For cultural accumulation, there has to be a reservoir of continuity. One brake on rapid change may be the process of cultural learning in the household, where initially the new generation learns from the parental one, so that by the time offspring are peer oriented perhaps the cultural core is already established. But this is hardly sufficient.

A second much more powerful force for conservatism is the power of systems. A single individual can change a word or two, but can hardly make much impact on a language. Even the greatest jurists have a barely discernable impact on a legal system, and the English kinship system has scarcely changed in half a millennium or more. Like machines, sociocultural systems resist random ad hoc changes because they have to function. Furthermore, unlike machines, sociocultural systems are the outcome of thousands of individuals each doing their part; my changing the script isn't going to stop the rest of you from carrying on. Given that sociocultural systems are, as it were, simply a web of concepts, intentions, and actions, their stubborn resistance to change is remarkable. Even when change appears on the surface, the system may remain stable. Changing hemlines, the pronunciations of words, or the style of pop songs (the kinds of examples that memetics focuses on) does not change the underlying system of clothing, language, or musical intervals. To borrow de Saussure's analogy for language, you can play chess with any arbitrary collection of shells or stones, providing the values of each piece are preserved. Systems are articulated sets of values and are not to be identified with the substances that temporarily substantiate them.

Cultures are systems larger than the individual. No one individual controls all the knowledge and practices, which are distributed throughout a population in a complex division of labor and cognition. Consider again something like a kinship system; the male and female perspectives on such a system differ in many essentials. There are interesting studies of distributed cognition (see Hutchins 1995) where a team (such as navigation team on a vessel) serves as a computing device to yield practical solutions (e.g., the course to navigate). If one asks what are the background conditions that make such a procedure possible, they include, first, the assumption of the fidelity of transmission of the procedure (we must think the others know what they are doing), second and crucially, the presumption of cooperation and trust—we must trust that we are each doing our bit as well as we can.

The general presumption of truth telling is essential to learning a language (how could you learn what "rabbit" means if the probability is that "That is a rabbit!" is false?). Cooperation and trust of this order are rare or non-existant in nonhuman animal behavior, and the evolutionary explanation of how it could arise runs straight into the conundrum of altruism: In an altruistic society the cheater always does best, so there are only limited, special circumstances in which altruism can be a stable evolutionary strategy (see Boehm, chapter 4 and Boyd and Richerson, chapter 5 this volume). One possibility is that it is this cooperation that has in evolution driven the mind reading that Tomasello (1999 and chapter 10) judges to be the crucial ingredient for culture. Being able to "read" others' intentions to a considerable depth is essential to cooperation, but it is also essential to detecting cheaters. A person might have all the right cooperative intentions up to some point, and then develop an additional intention to mislead. Consider. This example: When you and I cooperate to move a log, I have to see that you are intending to pick up your end if I pick up my end, but I also want to be assured that you don't later intend to make me drop it on my toe. Overall, the cognition of cooperation is much more demanding than the cognition of competition because competitive goals are preset but cooperative ones have to be communicated and established (see Levinson 1995 for the formal argument). The ability to mind read will act as some guarantee against cheating, or maximizing at the expense of one's fellows. Thus general considerations suggest that cognitive complexity may have been driven both by the cooperation that underlies culture and the need to protect it.

Notes

1. This assumes a life expectancy of 45 years for our prehistoric ancestors, with puberty at, say, 15 (as in current traditional societies, such as that is Rossel Island, discussed later).

2. Computer simulations show that a few hundred generations are easily sufficient to bring a rare mutant allele to dominance in a population, even with a low selection coefficient (see Durham 1991: 242–243 for references). In general, Lumsden and Wilson (1981) suggest a "thousand-year rule," under which 1000 years can be sufficient time for feedback from cultural to biological evolution. Dairying in the Near East certainly goes back 6000 years.

3. For some pithy remarks on Lumsden and Wilson's (1981) "culturgen" unit, see Hallpike (1982:13): "The definition is so vague . . . that a culturgen could be any discriminable aspect of human thought or behavior whatsoever: It is as though the "thing" were to be proposed as the basic unit of physics."

4. The ideational view of culture has a relatively short history and remains dominant only in American anthropology, being first articulated clearly by Goodenough (1957).

5. The modern human hand displays a range of grips that are quite impossible for an ape. *Homo habilis*'s more humanlike hands emerged with the first Oldowan tools more than two million years ago, but the size of the spinal chord remained limited right through into *Homo erectus*, implying relatively crude motor control. Even Neanderthals had hands that were distinctly different from those if modern humans (see Trinkaus 1992).

6. Oscine birds have song dialects, but only the form, not the function, is variable across dialects (see Hauser 1997: 273–300).

7. For the range of sound systems, see Maddieson (1984); for word order see Austin and Bresnan (1996); for noun versus verb, see Mithun (1999: 60–67).

8. Genetic drift increases over time at a rate that is inversely proportional to the size of the population; genetic distance d over time t in a population of size N increases according to the formula $d = t/2N$.

9. The measure of relative complexity in languages is nontrivial and has been much discussed, especially in the context of Creole languages. See the special issue of *Linguistic Typology* (2001, vol. 5 no. 2/3) for discussion and references.

10. As discussed in the first section, it is not necessary to assume that, say, witchcraft is an innate idea (then we all ought to subscribe to the identical doctrine) in order to account for its widespread belief. Human malevolence is empirically exemplified; it is the belief in a special kind of (supra-)natural force that has to be explained. And that might indeed be explained by our in-built social cognition, giving us a propensity to read the world as driven by intentions (see Levinson 1995).

11. Rossel islanders do, however, show significant proportions of markers on the Y-chromosome that are associated with Oceanic peoples (Manfred Keyser, personal communication).

12. In principle of course Rossel islanders could have borrowed fancy new names for good old, familiar objects, but it is unlikely. There are many different kinds of lexical borrowing, but in contexts of nonintense cultural contact (presumed relevant here) a major tendency is for lexical borrowings to be associated with new objects, practices, and concepts. In cases of intense contact, however, the bets are off; almost any aspect of language can be borrowed for a wide range of motives (see e.g., Thomason and Kaufman 1988).

13. The earliest (pre-Austronesian) archaeological evidence for the use of the betel nut goes back nearly 6000 years on the New Guinea mainland, but it does not appear in island deposits until perhaps 2000 years ago (see Kirch 1997: 40, 217).

References

Aoki, K., and Feldman, M. W. 1991. Recessive hereditary deafness, assortative mating, and persistence of a sign language. *Theoretical Population Biology* 39(3): 358–372.

Aiello, L. C., and Wheeler, P. 1995. The expensive tissue hypothesis: The brain and the digestive system in human and primate evolution. *Current Anthropology* 36: 199–221.

Armstrong, W. 1928. *Rossel Island.* Cambridge: Cambridge University Press.

Aunger, R. (ed). 2000. *Darwinizing culture: The status of memetics as a science.* Oxford: Oxford University Press.

Austin, P., and Bresnan, J. 1996. Non-configurationality in Australian Aboriginal languages. *Natural Language & Linguistic Theory* 14(2): 215–268.

Baldwin. J. M. 1896. A new factor in evolution. *American Naturalist* 30: 441–51.

Barkow, J. H., Cosmides, L., and Tooby, J. (eds.) 1992. *The adapted mind: Evolutionary psychology and the generation of culture.* Oxford: Oxford University Press.

Beise, J., and Voland, E. 2002. Differential infant mortality viewed from an evolutionary biological perspective. *History of the Family* 7: 515–526.

Berlin, B., and Kay, P. 1969. *Basic color terms: Their universality and evolution*. Berkeley, Calif.: University of California Press.

Boehm, C. 1996. Emergency decisions, cultural selection mechanics, and group selection. *Current Anthropology* 37: 763–793.

Boyd, R., and Richerson, P. J. 1985. *Culture and the evolutionary process*. Chicago: University of Chicago Press.

Brown, P. 2002. Language as a model for culture: Lessons from the cognitive sciences. In R. Fox and B. King (eds.), *Anthropology and culture* (pp. 169–192). Oxford: Berg.

Cavalli-Sforza, L. L., and Feldman, M. W. 1981. *Cultural transmission and evolution: A quantitative approach*. Monographs in population biology, 16. Princeton, N.J.: Princeton University Press.

Cavalli-Sforza, L. L., Menozzi, P., and Piazza, A. 1994. *The history and geography of human genes*. Princeton, N.J.: Princeton University Press.

Changeaux, J.-P. 1985. *Neuronal man*. New York: Oxford University Press.

Darwin, C. 1871. *The descent of man* (2nd Ed.). London: John Murray.

Darwin, C. 1872. *The origin of species by means of natural selection or the preservation of favoured races in the struggle for life*. (6th ed.). London: John Murray.

Dawkins, R. 1976. *The selfish gene*. Oxford: Oxford University Press.

Dawkins, R. 1983. *The extended phenotype*. Oxford: Oxford University Press.

Deacon, T. 1997. *The symbolic species*. New York: Norton.

Dennett, D. 1995. *Darwin's dangerous idea*. New York: Simon & Schuster.

Diamond, J. 1998. *Guns, germs and steel*. London: Vintage.

Diamond, J. 2002. Evolution, consequences and future of plant and animal domestication. *Nature* 418: 700–707.

Dobzhanzy, T. 1962. *Mankind evolving: The evolution of the human species*. New Haven, Conn.: Yale University Press.

Durham, W. 1991. *Coevolution*. Stanford, Calif.: Stanford University Press.

Edelman, G. 1987. *Neural Darwinism*. New York: Basic Books.

Elman, J., Bates, E., Johnson, M., Karmiloff-Smith, A., Parisi, D., and Plunkett, K. 1996. *Rethinking innateness*. Cambridge, Mass.: MIT Press.

Fox, R., and King, B. (eds.) 2002. *Anthropology and culture*. Oxford: Berg.

Gallistel, C. R. 1990. *The organization of learning*. Cambridge, Mass.: MIT Press.

Goodenough, W. 1957. Cultural anthropology and linguistics. In P. Garvin (ed.), *Report of the 7th annual round table meeting on linguistics and language study* (pp. 167–173). Washington, D.C.: Georgetown University Press.

Goody, J. 1977. *The domestication of the savage mind.* Cambridge: Cambridge University Press.

Gould, S. 1991. *Bully for Brontosaurus.* New York: Norton.

Gould, S., and Lewontin, R. 1979. The spandrels of San Marco and the Panglossian paradigm: A critique of the adaptationist programme. *Proceedings of the Royal Society of London* B205: 581–598.

Hallpike, C. R. 1982. The culturgen—Science or science-fiction. *Behavioral and Brain Sciences* 5(1): 12–13.

Hardin, C. L., and Maffi, L. 1997. *Color categories in thought and language.* Cambridge: Cambridge University Press.

Hauser, M. D. 1997. *The evolution of communication.* Cambridge, Mass.: MIT Press.

Humphrey, N. 1976. The social function of the intellect. Reprinted in R. Byrne and A. Whiten (eds.), *Machiavellian intelligence* (pp. 13–26). Oxford: Clarendon Press.

Hutchins, E. 1995. *Cognition in the wild.* Cambridge, Mass.: MIT Press.

Jackendoff, R. 1992. *Languages of the mind.* Cambridge, Mass.: MIT Press.

Jones, S., Martin, R., and Pilbeam, D. (eds.) 1992. *The Cambridge encyclopedia of human evolution.* Cambridge: Cambridge University Press.

Kirch, P. V. 1997. *The Lapita peoples: Ancestors of the Oceanic world.* Cambridge, Mass.: Blackwell.

Krings, M., Stone, A., Schmitz, R. W., Krainitzki, H., Stoneking, M., and Paabo, S. 1997. Neanderthal DNA sequences and the origin of modern humans. *Cell* 90: 19–30.

Kuhl, P. 1991. Perception, cognition and the ontogenetic and phylogenetic emergence of human speech. In S. E. Brauth, W. S. Hall, and R. J. Dooling (eds.), *Plasticity of development* (pp. 73–106). Cambridge, Mass.: MIT Press.

Kuhl, P. K., and Meltzoff, A. N. 1997. Evolution, nativism and learning in the development of language and speech. In M. Gopnik (ed.), *The inheritance and innateness of grammars* (pp. 7–44). New York: Oxford University Press.

Laland, K., and Brown, G. 2002. *Sense and nonsense: Evolutionary perspectives on human behaviour.* Oxford: Oxford University Press.

Laland, K., Odling-Smee, J., and Feldman, M. 2000. Niche construction, biological evolution, and cultural change. *Behavioral and Brain Sciences* 23:131–175.

Levinson, S. C. 1995. Interactional biases in human thinking. In E. Goody (ed.), *Social intelligence and interaction* (pp. 221–260). Cambridge: Cambridge University Press.

Levinson, S. C. 1997. Language and cognition: The cognitive consequences of spatial description in Guugu Yimithirr. *Journal of Linguistic Anthropology* 7(1): 98–131.

Levinson, S.C. 2000. Yélî Dnye and the theory of basic color terms. *Journal of Linguistic Anthropology* 10(1): 3–55.

Levinson, S. C. 2003. *Space in language and cognition.* Cambridge: Cambridge University Press.

Liep, J. 1983a. Ranked exchange in Yela (Rossel Island). In J. Leach and E. Leach (eds.), *The Kula new perspectives on Massim exchange* (pp. 503–524). Cambridge: Cambridge University Press.

Liep, J. 1983b. This civilizing influence: The colonial transformation of Rossel Island society. *Journal of Pacific History* 18: 113–131.

Lumsden, C. J., and Wilson, E. O. 1981. *Genes, mind and culture.* Cambridge, Mass.: Harvard University Press.

Lynch, J., Ross, M., and Crowley, T. (eds.) 2002. *The Oceanic languages.* London: Curzon.

Maddieson, I. 1984. *Patterns of sounds.* Cambridge: Cambridge University Press.

Margulis, L. 1981. *Symbiosis in cell evolution.* New York: Freeman.

Mayr, E., and Diamond, J. 2001. *The birds of Northern Melanesia: Speciation, ecology, and biogeography.* Oxford: Oxford University Press.

McGrew, W. C. 1992. *Chimpanzee material culture: Implications for human evolution.* Cambridge: Cambridge University Press.

Midgley, M. 2001. Why memes? In H. Rose and S. Rose (eds.), *Alas poor Darwin: Arguments against evolutionary psychology* (pp. 67–84). London: Vintage.

Mithun, M. 1999. *The languages of Native North America.* Cambridge: Cambridge University Press.

Paracer, S. 1986. *Symbiosis.* Hanover, N.H.: University of New England Press.

Pinker, S. 1994. *The language instinct.* New York: William Morrow.

Plotkin, H. 1997. *Evolution in mind: An introduction to evolutionary psychology.* New York: Penguin.

Rendell, L., and Whitehead, H. 2001. Culture in whales and dolphins. *Behavioral and Brain Sciences* 24: 309–382.

Rose, S. 1997. *Lifeliness: Biology, freedom, determinism.* Harmondsworth, UK: Allen Lane.

Rose, H., and Rose, S. (eds.) 2001. *Alas poor Darwin: Arguments against evolutionary psychology.* London: Vintage.

Spriggs, M. 1997. *The island Melanesians.* Malden, Mass.: Blackwell.

Sterelny, K. 2001. *Dawkins vs. Gould: Survival of the fittest.* Cambridge: Icon Books.

Talmy, L. 2000. *Toward a cognitive semantics* (vols. 1 and 2). Cambridge, Mass.: MIT Press.

Terrill, A., Reesink, G., and Dunn, M. 2002. The East Papuan languages: A preliminary typological appraisal. *Oceanic Linguistics* 41(1): 28–62.

Thomason, S., and Kaufman, T. 1988. *Language contact, Creolization, and genetic linguistics.* Berkeley, Calif.: University of California Press.

Tomasello, M. 1999. *The cultural origins of human cognition.* Cambridge, Mass.: Harvard University Press.

Trinkaus, E. 1992. Evolution of human manipulation. In S. Jones, R. Martin, and D. Pilbeam (eds.), *The Cambridge encyclopedia of human evolution* (pp. 346–349). Cambridge: Cambridge University Press.

Voland, E. 1998. Evolutionary ecology of human reproduction. *Annual Review of Anthropology* 27: 347–374.

Whiten, A., Goodall, J., McGrew, W. C., Nishida, T., Reynolds, V., Sugiyama, Y., Tutin, C. E. G., Wrangham, R., and Boesch, C. 1999. Cultures in chimpanzees. *Nature* 399: 682–685.

Williams, G. C. 1966. *Adaptation and natural selection: A critique of some current evolutionary thought.* Princeton, N.H.: Princeton University Press.

Wilson, D. S., and Sober, E. 1994. Re-introducing group selection to the human behavioral sciences. *Behavioral and Brain Sciences* 17(4): 585–654.

I Emergence of Culture in Evolution

2 Quantum Leaps in Evolution

Claude Combes

Nature does make jumps
—Thomas H. Huxley, The origin of species, reprinted in D. L. Hull, (ed.) *Darwin and His Critics*

Genetic Systems

In the course of evolution, genetic information has been split into millions of different entities (the species), isolated from each other in a way that justifies the representation of the history of life as an enormous tree whose branches seem to diverge indefinitely. In reality, genetic information is not necessarily isolated forever. Certain branches of the tree, sharing sometimes only a very distant common ancestor, can also converge and combine the information that they carry with them.

Such associations are systems, in the sense that the properties of a system formed by two or more elements are different from the simple addition of the properties of its elements. Systems in which partners are supposed to cooperate are called mutualistic, while systems in which relationships are supposedly conflictual are called parasitic.

Mutualism Distinguished from Parasitism
It is generally assumed that selective pressures are different in parasitic and mutualistic systems. For a system to exist, its elements must come in contact with each other (encounter) and be capable of living together (compatibility). Encounters and compatibility between two different living species can be symbolized by two "filters" or "irises" (Combes 1997, 2001), whose degrees of opening are determined by the genes of both partners. In parasitism, the degree of opening of the filters is the result of two opposing forces; the fitness of one partner (the parasite) is increased when the filters are opened, whereas the fitness of the second partner (the host) is maximized when the filters are closed. In mutualism, selection in both partners tends to open the filters; any genes or gene combinations from either of the species that open the filters are at a selective advantage (at least to the point where some regulation can be necessary).

However, the distinction between parasitism and mutualism is not always clear. A first difficulty is that even in the associations that seem to be the most clearly mutualistic, conflicts may still exist between the partners, as is the case between the eukaryotic cell and its mitochondria (see Couvet et al. 1990; Hurst et al. 1996; Herre et al. 1999), proving that each partner remains fundamentally selfish.

A second difficulty is the evaluation of costs and benefits in the system. Smith (1992), for instance, regards mutualism more as an exploitation of the parasite by the host than as a true two-sided beneficial system. "The contrast between symbiosis and parasitism is a contrast between hosts which exploit their associates, and hosts which are being exploited by their associates" (Smith 1992: 4) The manner in which heterotrophic organisms become associated with autotrophic ones seems to support Smith. There are associations of nonautotrophic cells with autotrophic bacteria (which become chloroplasts), associations of nonautotrophic cells with cells harboring autotrophic bacteria (chloroplasts), and conservation of autotrophic bacteria (chloroplasts) by nonautotrophic cells after digestion of autotrophic prey (figure 2.1). These different types of associations (which may be called primary mutualism, secondary mutualism, and second-hand mutualism, respectively) are convincing evidence that heterotrophic organisms are parasites rather than mutualists of their autotrophic associates.

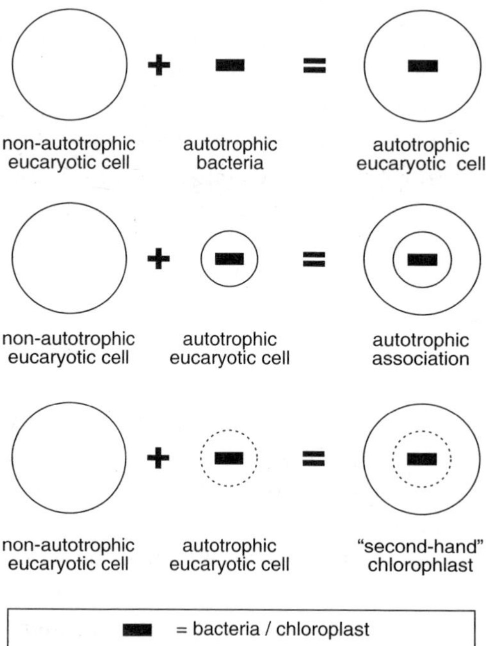

Figure 2.1

Three different ways in which heterotrophic organisms become associated with autotrophic ones.

A third difficulty is that certain associations are on the frontier between parasitism and mutualism. This is illustrated by the mite *Unionicola ypsiliphora*, which parasitizes freshwater mussels and whose immature forms leave the bivalves in search of the aquatic pupae of chironomids. Upon encountering a pupa, the mite infects the adult insect when it emerges and then gorges itself with its host's hemolymph. The mites thus obtain both a meal and a means of transport to a new aquatic habitat. The mites, which are bright red, infect both sexes of the chironomid. However, after mating, only the female chironomids fly off in search of an aquatic site for ovipositing. Thus, only those mites attached to females should logically return to water and survive. The mites overcome this difficulty by transferring from males to females during copulation. At this stage of the observations, *U. ypsilifora* clearly seems to be a parasite because it is expected to be detrimental for both parasitized males and females. However, McLachlan (1999) has demonstrated that infection improves the mating success of the males. While the ratio of infected males in the whole population is about 4 percent, it increases to 15 percent in mated pairs (figure 2.2). A possible explanation is that females prefer males that exhibit the red color of the parasites; another is that parasites "manipulate" male behavior. From the male host's perspective, the carrying

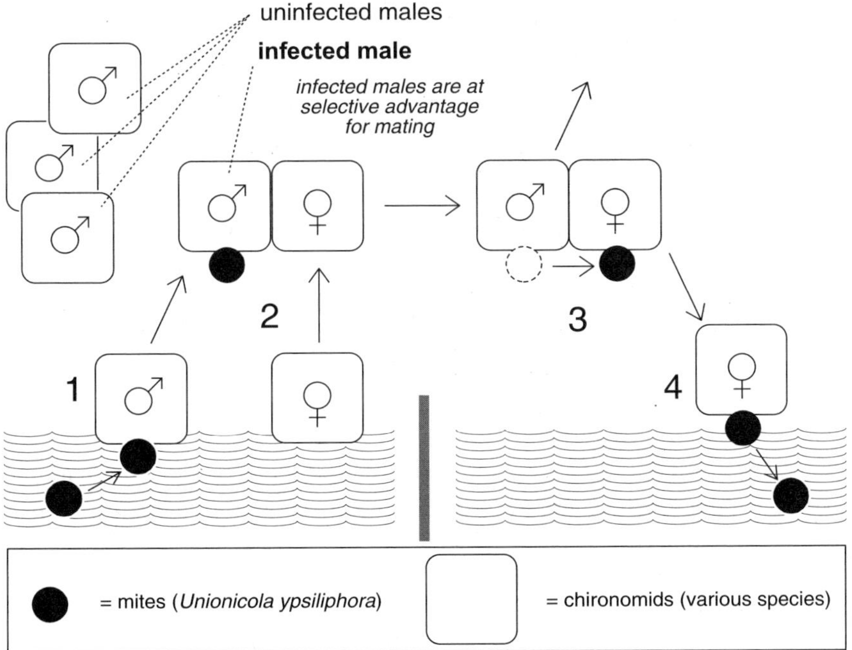

Figure 2.2
Mating in the mite *Unionicola ypsiliphora*. (Based on data from McLachlan 1999.) Direct infection of females is not represented.

of parasites means an increased probability of transmitting its genes in such a way that *U. ypsiliphora* is a parasite of female chironomids, but a mutualist for males.

What Gives the System New Properties?

In parasitism as well as in mutualism, two different genomes remain close to each other during a certain duration of time and can thus interact. This long-lasting interaction (Combes 2001) is different from what happens in a predator–prey interaction, in which the information carried by one of the entities is instantaneously destroyed.

A long-lasting interaction between the genetic information of two entities gives new properties to the system. A first new property is the extended phenotype, which results from the expression of one genome into the phenotype of the other (Dawkins 1982). For instance, some parasites manipulate the appearance and/or behavior of their intermediate hosts to make them more conspicuous to their predators, which are the definitive hosts (Moore 1995; Combes 2001).

A second new property can appear in a mutualistic system if genes belonging to two different genomes cooperate to construct what I call here a coordinated phenotype. A good example of a coordinated phenotype is provided by the luminous organs of deep-sea fishes. Bioluminescence is common in deep-sea fishes since sunlight is sparse at 300 m and absent further down. Some of these fish can produce light themselves because they have the genes coding for the substrates and enzymes of the luciferin–luciferase complex. Others, however, emit light only because of an association with bacteria of the genera *Photobacterium* or *Vibrio*. Ponyfish harbor the bacteria in their esophagus and build a true lighthouse around this light source. The muscles lining the esophagus have become transparent; the swim bladder has been transformed into a reflector; and chromatophores are used as interrupters (see MacFall-Ngai 1989). The lighthouse is thus composed of parts coded by the bacterial genome and parts coded by the fish genome (figure 2.3). In such an association, mutualists are equivalent to enormous insertions of genomic information. This explains why, from an evolutionary point of view, a mutualistic system is regarded as a single unit by selection (see Maynard Smith 1989a, b).

A third new property occurs in certain mutualistic systems when the long-lasting interaction results in an exchange of the supports of information themselves, i.e., fragments of DNA. In the previous example, for instance, one can imagine that the five genes necessary to produce light are transferred from the bacterial genome to the fish genome. This would obviate the need to reacquire luminous bacteria in each new generation of fish. It is not clear whether this kind of event has occurred in certain deep-sea fish. Nevertheless, gene transfers are well documented in other systems; for instance, between the nuclear genome of the eukaryotic cell and the DNA from its mitochondria or chloroplasts. Mobile elements could play an important role in gene transfers. It is likely that gene transfers have important consequences in terms of

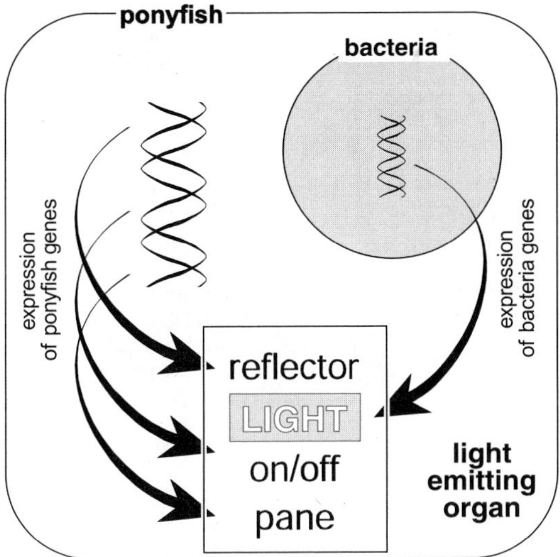

Figure 2.3
An example of coordinated phenotype: the light-emitting organ of ponyfish.

gene expression and regulation. The new combinations are metagenomes (Combes 2001).

Mutualism and the Tempo of Evolution

In recent years, several papers have emphasized the fact that the tempo of evolution is certainly less "regular" than was thought 50 years ago. One reason is that a more important role is now attributed to rapid events related, for instance, to tectonic or climatic changes or the isolation of small populations (see Carroll 2000; Gould 2002). Another is that the importance of what Maynard Smith and Szathmary (1995) call "major transitions" has been recognized. Major transitions refer to mutualistic associations between organisms coded by different genomes. By acquiring innovative genes (Combes 2001) as a set, instead of by the long process of many events of mutation and selection, an individual may have a strong selective advantage over other individuals in a population. It is the occurrence of mutualistic systems that has permitted a series of quantum leaps in the course of evolution. The eukaryotic cell is the classic example of such a successful association.

I have written earlier here that genetic information has been split into many different entities during the course of evolution. The same can be said of cultural information in the course of human evolution (figure 2.4). Cultures "speciated" in the same way genomes did, in relation to isolating mechanisms. This is not surprising because

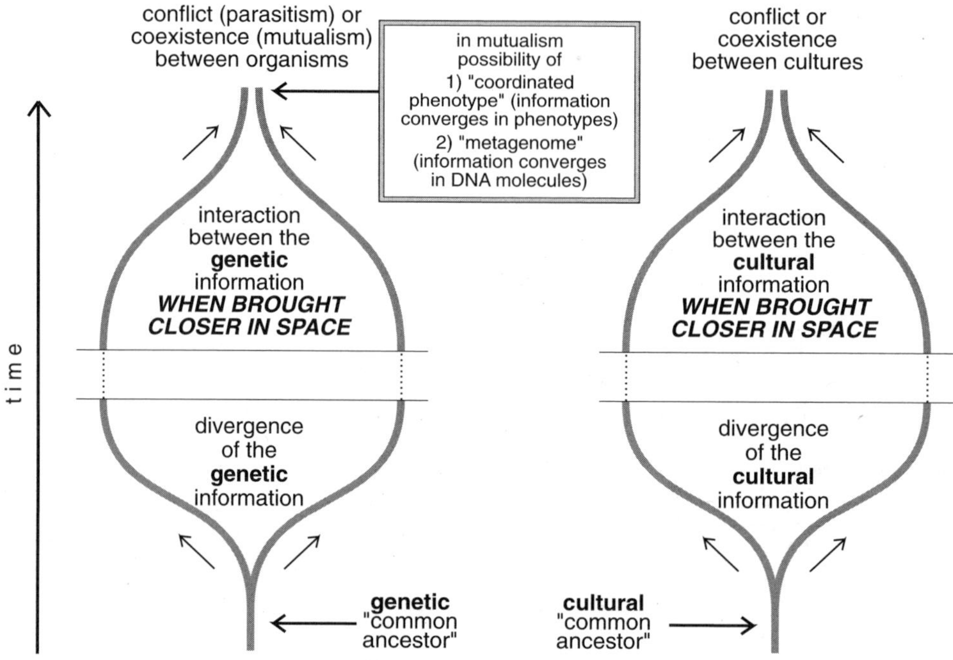

Figure 2.4

Processes of divergence and convergence of information during evolution.

a genome and a culture are two forms of information (see Feldman and Laland 1996) for which only the mode of transmission is different—Darwinian versus Lamarckian. In the Darwinian mode, all information gathered by somatic cells is lost at each generation, whereas in the Lamarckian mode it can be transmitted, at least in part.[1] When two different entities of cultural information happen to form a system, it seems possible to use the concept of encounter and compatibility filters as we did for genetic information. An encounter depends on whether or not human populations come in contact; compatibility depends on whether they become conflictual or enrich each other. When cultures from different origins meet somewhere on our planet (and sometimes perhaps cultural traits of single humans in the same population), either one dominates the other or the encounter leads to a major transition, an alternative which is, paradoxically, similar to a Darwinian process.

A tentative conclusion is that in cultural traits as well as in genotypes, divergence is essential to provide diversity, but convergence is necessary to determine evolutionary quantum leaps. In other words, the moments when symbioses of information (genetic + genetic, cultural + cultural, and possibly genetic + cultural) are being formed are milestones for evolution, from the first living molecule that appeared 3.5 billions years ago to the modern human society.

Note

1. Let us remark, however, that somatic cell death is considered itself to have been "a major prerequistite for the development of multicellular organisms" (Rudel and Meyer 1999). These authors consider that the ancestral mitochondria were able to kill the host cell and that it was only later that the host cell could control the killing mechanisms of the mitochondria and from this generate the capacity for apoptosis.

References

Carroll, R. L. 2000. Toward a new evolutionary synthesis. *Trends in Ecology and Evolution* 15: 27–32.

Combes, C. 1997. Fitness of parasites. Pathology and selection. *International Journal for Parasitology* 27: 1–10.

Combes, C. 2001. *Parasitism. The ecology and evolution of intimate interactions.* Chicago: The University of Chicago Press.

Couvet, D., Atlan, A., Belhassen, E., Gliddon, C., Gouyon, P. H., and Kjellberg, F. 1990. Co-evolution between two symbionts: The case of cytoplasmic male-sterility in higher plants. *Oxford Surveys in Evolutionary Biology* 7: 225–249.

Dawkins, R. 1982. *The extended phenotype.* Oxford: Oxford University Press.

Feldman, M. W., and Laland, K. N. 1996. Gene-culture coevolutionary theory. *Trends in Ecology and Evolution* 11: 453–457.

Gould, S. J. 2002. *The Structure of Evolutionary Theory.* Cambridge, Mass.: Harvard University Press.

Herre, E. A., Knowlton, N., Mueller, U. G., and Rehner, S. A. 1999. The evolution of mutualisms: Exploring the paths between conflict and cooperation. *Trends in Ecology and Evolution* 14: 49–53.

Hurst, L. D., Atlan, A., and Bengtsson, B. O. 1996. Genetic conflicts. *Quarterly Review of Biology* 71: 317–364.

MacFall-Ngai, M. J. 1989. Luminous bacterial symbiosis in fish evolution: Adaptive radiation among the leiognathid fishes. In L. Margulis and R. Fester (eds.), *Symbiosis as a source of evolutionary innovation* (pp. 381–409). Cambridge, Mass: MIT Press.

Maynard Smith, J. 1989a. A darwinian view of symbiosis. In L. Margulis and R. Fester (eds.), *Symbiosis as a source of evolutionary innovation* (pp. 26–39). Cambridge, Mass: MIT Press.

Maynard Smith, J. 1989b. Generating novelty by symbiosis. *Nature* 341: 284–285.

Maynard Smith, J., and Szathmary, E. 1995. *The major transitions of evolution*. New York: W. H. Freeman/ Spektrum.

McLachlan, A. 1999. Parasites promote making success: The case of a midge and a mite. *Animal Behaviour* 57: 1199–1205.

Moore J. 1995. The behavior of parasitized animals. When an ant … is not an ant. *Bioscience* 45: 89–96.

Rudel, T., and Meyer, T. F. 1999. Infection of human cells by *Neisseria*—A paradigm of ancestral apoptosis? *Nova Acta Leopoldina* 307: 71–86.

Smith, D. C. 1992. The symbiotic condition. *Symbiosis* 14: 3–15.

3   The Emergence of Culture in the Context of Hominin Evolutionary Patterns

Robert A. Foley

The evolution of culture remains the most challenging problem in the human sciences. If one accepts that human capacities must have an evolutionary basis, then how can one explain the transformation from a species that lacked cultural capabilities to one that possessed them without assuming the presence of those characteristics in the first place? This difficulty has pushed research in two different directions. On the one hand there are those who would say that there can be no continuity between the "acultural" and the "cultural," except inso far as there is a saltational leap from the one to other, a leap shrouded in mystery. Continuity does not exist in this model (see Premack and Hauser in chapter 13, for example). On the other hand, the origins of culture in the human lineage can be explained by its presence in other lineages; in other words, we have it because chimpanzees have it. This is the approach adopted most strongly by McGrew (1992), but also inherent in the view that chimpanzees exhibit cultural variation (Whiten et al. 1999). However, while assigning culture to chimpanzees may make the problem of continuity less pressing in terms of human evolution, it merely takes it farther back phylogenetically.

The problem these approaches face is that of synchronism. Effectively the comparison between a cultural and noncultural being, whether it be humans and chimpanzees or chimpanzees and gorillas, is between two living species, and ones that cannot be ancestral to each other. The continuity or discontinuity of culture as an evolutionary outcome is in the end dependent upon the interpretation placed on chimpanzee behavior and the acquisition of new data on their abilities (Foley 1991).

The way out of this "chimpanzee tyranny" is to add a directly temporal and historical element; in other words, to look at the actual evolution of more and more humanlike capacities during the course of evolution, especially since the split with the chimpanzees. However, any such paleobiological approach is strongly constrained by the very limited nature of the fossil and archaeological evidence. The only sources of information are crude estimates of brain size, inferences from brain structure, correlations with anatomical structures such as the larynx, and information derived from technology or site structure. Despite the inherent limitations of these data, it is this perspective that I pursue here. In particular I address four questions concerning the origins of culture which are susceptible to this approach: (1) Was there a cultural revolution? (2) What is the pattern of cultural preconditions? (3) What was the cultural status of the different hominin taxa? (4) What was happening as cultural properties evolved? First, however, it is necessary to consider briefly the question of what culture may mean in this sort of approach.

Culture and Evolution

Culture can be defined in numerous ways, most of which are overlapping and open to criticism. Broadly speaking, culture can either be thought of as a series of end products that arise from the inherent nature of humans and the way in which they interact with their natural and social environment, or else as the process that produces these outcomes (Cavalli-Sforza and Feldman 1981). This latter approach is more strongly supported today, and emphasizes the cognitive foundations of cultural properties (Tomasello 2000). Culture is a way of doing things rather than the things themselves. For example, both humans and birds make "houses," but the way in which humans perform this activity is considered to be cultural, whereas that of birds is instinctive, or else is learned in a relatively simple way. Here I assume that it is this cognitive meaning of culture that requires an evolutionary explanation. However, the cognitive foundations of culture cannot be observed directly, especially in the fossil and archaeological record, and so it is the fact that this cognitive capability is either correlated with anatomical preconditions (e.g., brain size and structure), or else leads to certain outcomes (e.g., ethnic variability) that provides us with access to information about cultural evolution.

It should be stressed that this approach to cultural evolution is a partial one. In the end, culture can only be understood as a cognitive process (e.g., Tomasello in chapter 10), or as a neurobiological one (Singer in chapter 9), existing in particular social contexts (Levinson in chapter 1), and leading to new evolutionary patterns and processes (Boyd and Richerson in chapter 5). What paleobiology can add is some sense of the context in which these occurred, and hence perhaps examine why it might have evolved, and the time depth involved.

The correlates of a more cultural form of cognition, it should be stressed, are general ones, and can only be inferred in a relatively unrefined way on the basis of archaeological information. However, at the core of most definitions of culture lie three major components. One of these is associated with learning; that is, the ability to acquire new information independent of any genetic basis, and more subtly, the depth and extent of that learning (i.e., its complexity). The second consists of a group of attributes associated with social organization and structure; these would generally come under the heading of social complexity, but can in turn be broken down into mech-

Table 3.1

Relation between cultural components and the archaeological and fossil record

Broad correlative components of culture	Potential paleobiological manifestations
Learning capacity	Technology and technological variation Brain size?
Social organization and structure	Archaeological density, structure, and distribution Sexual dimorphism in fossil hominins Non-ecologically functional elements of material culture
Traits associated with symbolic thought	Brain size? Anatomical basis for language Variation in material culture
Maintenance and change of tradition	Regional variation and longevity of archaeological components

anisms for maintaining intragroup relationships and alliances, both within and between sexes; systems for parental investment, care, and control; and mechanisms for regulating relationships between communities. The third component includes traits associated more directly with symbolic thought. Within this component would be symbolic expression itself, both verbally (language) and using other means (material culture), as well as systems of belief that would extend to religion and other emotionally based belief systems.

In addition to these three core components of culture, it would also be necessary to consider the persistence and diversity of tradition, but these are in some sense outcomes of the other components that relate to the mutability of the means by which they are replicated and modified. Thus, when the emergence of culture is viewed here from a paleobiological perspective, it is these features that are considered to be the emergent properties of the underlying cognition. Table 3.1 shows the associations among these three components and their potential paleobiological and archaeological manifestations.

One further consideration of how cultural evolution can be investigated from an evolutionary point of view needs to be stressed. It is clear that within anthropology, culture is treated very much as a total package and largely as an emergent property of the capacities of the human species. On this basis it is used as an analytical category in its own right; humans are cultural and other species are not, or humans have culture and other species do not. This lays the groundwork for the debates that occur in primatology, psychology, and anthropology over whether chimpanzees (to take the primate example) do or do not have culture or cultural capacities. However, this is a function of the synchronic approach described earlier that looks at either humans or other species living today.

Because evolution occurs through time, any lineage must pass from a state where certain traits are absent to a state where they are present. This is true for both behavioral and morphological characteristics. Although fossil hominins overall represent an adaptive array rather than a single lineage, nonetheless they also reflect an evolving trend, and new traits can be seen occurring at various points along this trend. Cultural abilities must also have appeared in the same way, and there are two models for how this may have occurred. One model is that they emerged as a single package in one event. If this were the case, then it would be appropriate to use the culture concept in paleobiology in the same way it is used in the more synchronic disciplines. However, it is also possible—indeed, I would argue a strong probability—that the components of culture evolved incrementally, with different elements coming into play at different times (Foley 1991, 1996). If this is the case, then the use of culture as an analytical concept is inappropriate, and instead it is necessary to break it down into its components and seek to understand their emergence in a more reductionist manner. It is this approach that is adopted here. It is perhaps a less theoretically exciting path of investigation, but it has the advantage of making fewer assumptions and being more susceptible to empirical testing.

Was There a Cultural Revolution?

The main area of controversy in human evolution in recent years has been the origins of modern humans. There has been a growing consensus that most of the available evidence—genetic, paleontological, and archaeological—supports a model in which modern humans evolved relatively recently, that is, less than 200,000 years ago (Stringer and Andrews 1988; Goldstein et al. 1995; Hammer 1995; Lahr and Foley 1998; Ingman et al. 2000; Klein 2000), and that the ancestors of all living humans are descended from a small population (less than 50,000, and probably much smaller) that lived in Africa (Harpending et al. 1993). From this evidence it has been inferred that modern humans dispersed from 100,000 years ago from Africa to other parts of the world, and in doing so displaced nonmodern hominins, which thus became extinct (Stringer 1989; Lahr and Foley 1994). This in turn led to increasing interest in establishing the cognitive and behavioral differences between modern humans and other hominins, especially Neanderthals, on the grounds that there must have been significant differences to account for the success of the former and the extinction of the latter. From this came what can broadly be considered the cultural or symbolic revolution model; modern humans had some cognitive and behavioral capabilities that were absent in other hominins.

This model has a variety of forms and explanatory elements, but broadly speaking, they share the view that the Upper Paleolithic, which occurs around 45,000 years ago in western Eurasia, is evidence of a major shift in human cognitive and cultural abilities, and that it is associated with one of a number of major new evolutionary acquisitions: language according to Davidson and Noble (1991), symbolic thought according to Knight (Knight et al. 1995) and Klein (1995), and the reorganization of cognitive modules according to Mithen (1996). The evidence for this is associated with the emergence of more complex and variable technology, the appearance of art, and the growing evidence for ethnic differentiation among human populations, expressed in material culture. In other words, they all see a cognitive revolution occurring about 50,000 years ago, linked to the evolution of modern human symbolic and linguistic capacities—the symbolic revolution. To some extent this has also been equated with the origins of full cultural capacities.

One model for the evolution of human culture is therefore that although there were preconditions present in earlier hominins, it actually takes its full modern form in a relatively short evolutionary change that occurred within the past 100,000 years, based on the emergence of human symbolic thought. How solid is the evidence for this model? I would argue that the evidence is not that secure, and that a more gradual model fits the available data rather better.

Firstly, the chronological association between the genetic bottleneck, modern anatomy, and behavior is not that strong. Figure 3.1 shows the association between modern humans defined anatomically (AMH) and the presence of blade technologies (mode 4 technologies, marked with a star), which according to some is one of the signals of modern human behavior. It can be seen that only in Europe and northern Africa is there a good relationship. In Africa, modern human anatomy precedes the appearance of blade technology by many millennia. Where there are early blade technologies, these actually are relatively ephemeral and disappear—e.g., the Howieson's Poort in southern Africa (Deacon 1992) and the pre-Aurignacian in North Africa (McBurney 1967). Although there is some evidence for other aspects of modern human behavior, for example, engravings, shell beads, the use of ochre and bone tools, among the early African modern humans (Knight et al. 1995; Henshilwood and Sealy 1997; Henshilwood et al. 2004), this can also be found in older contexts as well (McBrearty and Brooks 2000). These all indicate either that behavioral modernity is considerably older, or that different elements evolved independently over a considerable time.

Essentially the Upper Paleolithic appears to be too late and too regionalized to be a marker of modern human cultural capacities that are based on a biological propensity for culture (Foley and Lahr 1997). The genetic evidence adds another component to this argument. All the available data suggest that modern human populations today have the same capacities for culture, for language, and for symbolic thought, the main traits considered to be distinctly human. These may be considered species-specific universals. However, by 45,000, the date of the Upper Paleolithic, the genetic evidence suggests that the major branches of human populations had already diverged,

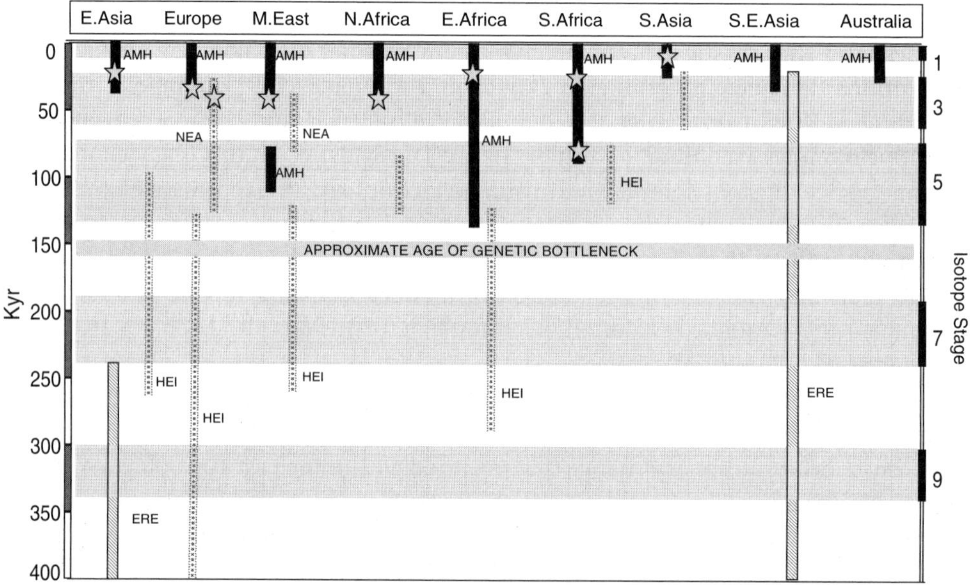

Figure 3.1

Association among the genetic bottleneck, hominin taxa, and the appearance of blade technology (mode 4) in the later Pleistocene. Only in Eurasia does blade technology appear with anatomically modern humans. In Africa there are earlier appearances, but these are ephemeral, and most early modern humans are associated with mode 3 technologies (Middle Stone Age/Middle Paleolithic). Key: AMU, anatomically modern humans; NEA, *neanderthalenses*; HEI, *heidelbergensis*; ERE, *erectus*; *, mode 4 technologies. (Adapted from Foley and Lahr 1997.) Shaded phases indicate warmer periods during the glacial cycles.

particularly the deeper African clades and Australians (Vigilant et al. 1991). If a major biological species-specific trait evolved as late as 50,000 years ago, it would either be absent in some human populations now, or else there would have had to have been a selective sweep across the global population at this time—and there is no genetic evidence for this event.

There is a further problem with the symbolic revolution model, and this relates to the shared abilities of modern humans and Neanderthals. There are a number of contrasts between Neanderthals and modern humans overall, especially if the comparison is made between Holocene AMH and Neanderthals, and there are certainly some comparisons that can be made between them and the modern humans of the late Pleistocene, the cave art of Lascaux being the supreme example. However, a comparison across the earlier part of the history of modern humans and Neanderthals, i.e., when they were directly contemporary between, say, 130,000 and 50,000 years ago,

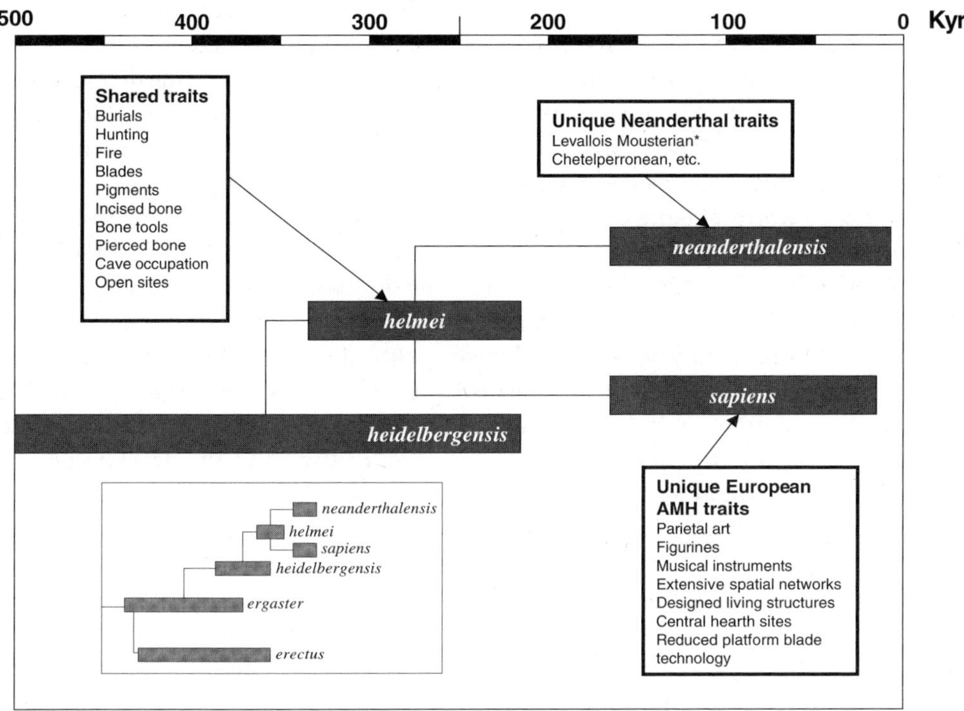

Figure 3.2

Phylogenetic relationships of *Homo* (inset) and of later *Homo* in particular, showing shared and unique behavioral characteristics of Neanderthals and modern humans.

is less striking (see figure 3.2). Both species use the same basic technology—mode 3 prepared core or Middle Stone Age/Paleolithic (Foley and Lahr 1997); both bury their dead (Valladas et al. 1987) and exhibit about the same level of regional and chronological variation. They are clearly both proficient hunters (Stiner et al. 1999). At a later time Neanderthals also perhaps showed some ability to make simple art objects (Hublin et al. 1996). Finally, the anatomical evidence would suggest that they were both capable of language (McLarnon 1996). These similarities should perhaps not be a surprise, because brain size, if that is an indication, is much the same in both species and is considerably larger than that found in any other hominin taxon (Aiello and Dunbar 1993).

The implication of these data is that while there were undoubtedly differences between Neanderthals and modern humans, the contrast is perhaps not as great as has often been thought. Perhaps a more important point to make is that because modern humans and Neanderthals are sister clades, we would expect their behavior—and by implication their cognition—to be closer to each other than either is to any other fossil hominin. In most cognitive tests it would be a reasonable guess that Neanderthals and modern humans would both perform considerably better than *H. erectus* or *H. heidelbergensis*. In this context, the transition from the Acheulean industries (mode 2) to the prepared core techniques of the Middle Stone Age/Paleolithic (mode 3) may have been of greater significance in cognitive and cultural terms than the change from mode 3 to mode 4 (Upper Paleolithic/Later Stone Age).

It is perhaps worth making one further point on this subject, and that is that when we look at the cultural complexity of modern humans today, it is clear that there has been a considerable increase in complexity over the past 20,000 years. This is most noticeable in the scale of human societies, the complexity of social institutions, the development of sedentary societies and larger political organizations, and the elaboration of material culture. What this means is that the Upper Paleolithic may mark a significant transformation in human cultural evolution, as was the transition to mode 3, but it is probably the case that the origins of agriculture and the development of sedentary societies represented as great a transition, and thus that the development of cultural complexity neither started nor finished with the origins of modern humans, but was a more drawn-out process (Foley 2001).

In summary, the answer to the question of whether there was a symbolic revolution is almost certainly "no"—at least not a major biologically based one at the time of the Upper Paleolithic. This is too late and too regional. Furthermore, Neanderthals share many derived traits with modern humans, which suggests that these traits may have been present in a common ancestor some 300,000 years ago or more. Finally, the scale of changes associated with the origins of modern humans is not significantly larger than changes seen earlier or later, and thus should be seen as part of a larger pattern.

What Is the Pattern of Cultural Preconditions Among Earlier Hominins?

If cultural capacities do not appear in a single revolutionary pattern in the Upper Pleistocene, it is obvious that the next question is, what can we say about the cultural preconditions that may have been present in earlier hominins, and that can be used to reconstruct the sequence of events that led to fully modern cultural behavior? The evidence for exploring this issue comes from the morphological information that can be derived from the hominin fossils and from the archaeological record for aspects of behavior.

The first point to make is that there is considerable diversity among hominins, so that it is not possible to see them as a single lineage evolving through anagenesis (Foley 1998; Tattersall 2000). At virtually all times in the past 5.0 million years there is evidence of the existence of more than one hominin taxon. This means that we must treat these taxa as distinct lineages and species, rather than as gradations along a continuum from ape to human. From these fossils we can derive a number of observations and measurements on the basis of which it is possible to make inferences about their potential for "culture" and their cognitive propensities. It should be stressed that these inferences cannot be about particular abilities, but about the relative status of each taxon in relation to both chimpanzees and humans, the comparative framework available to us. They will not tell us what hominins were capable of precisely, but can perhaps pinpoint when key shifts may have occurred in overall cognitive potential.

The problem, however, is deciding what characteristics we need to look at to make inferences about the cognitive abilities of extinct hominins that might be of relevance for their cultural abilities. The primary possibilities are brain size, life history parameters, and technology. Other aspects that will not be considered here might include the range of habitats occupied and ecological strategies pursued (evidence for cultural flexibility), size and density of archaeological sites (group size and complexity), and degree of sexual dimorphism (social behavior). It is essential that whatever trait is chosen can be viewed on a quantitative scale that can be used to make comparisons across taxa (Foley 1996).

Brain size has been the main source of information about the relative cognitive abilities of fossil hominins (Jerison 1973; Martin 1983). While there has been some discussion relating to brain structure, fossil endocasts are rare and have proved to be relatively intractable to much of this type of inference (Holloway 1983). Figure 3.3

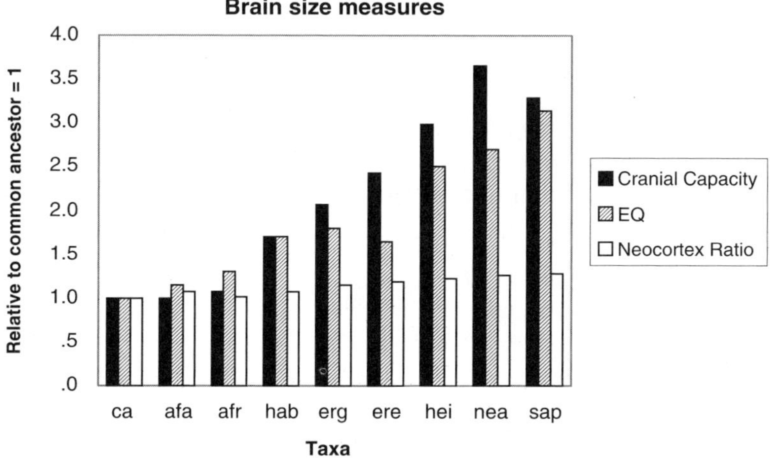

Figure 3.3

Measures of brain size for extinct hominin taxa, all espressed as a ratio to the hypothesized common ancestor. Cranial capacity gives absolute size; the EQ (encephalization quotient) gives relative size based on allometric relationships among mammals; and the neocortex ratio is the ratio of the neocortex to the rest of the brain. See Foley (1996) for details. Key: ca, common ancestor; afa, *afarensis*; hab, *habilis*; erg, *ergaster*; ere, *erectus*; hei, *heidelbergensis*; nea, *neanderthalensis*; sap = *sapiens*.

shows a number of brain size measures for hominins. These are not standard measures. First, three variables are shown: the actual brain size in terms of cranial capacity; the encephalization quotient (EQ), which is corrected for body mass (because body size is a major determinant of brain size across taxa) (Jerison 1973; Martin 1981); and neocortex ratio [the size of the neocortex relative to the brain as a whole, which Dunbar (1992) has argued is a better reflection of the changes in brain size that are related to higher cognitive function]. Second, these values have been scaled relative to a common ancestor, so that for each variable the common ancestor (assumed to be close to the value of a chimpanzee) is 1. By putting the data in this format, it is possible to see how trends in each variable might compare (Foley 1996).

Patterns of growth can also be used to some extent to make inferences about the cognitive and behavioral attributes of extinct hominins. The logic behind this statement is that it has been observed comparatively that there is an association among primates for a relationship between brain size, which is related to behavioral capacity, and delayed maturation and longevity (Harvey and Clutton-Brock 1985; Harvey et al. 1987, 1989; Harvey and Pagel 1988, 1992). Animals that mature more slowly have greater scope and potential for learning, and also require a greater degree of parental care, and hence social interactions (Clutton-Brock 1991). Humans have a particularly delayed development, and the brain only reaches its adult size at about the age of 6 years, compared with 3 years in chimpanzees. By comparing absolute dental age, based on perikymata and enamel striata, with stage of biological maturation, based on stage of tooth eruption, it is possible to estimate the change in life history pattern for fossil species (Bromage and Dean 1985; Dean et al. 1986; Beynon and Dean 1988; Smith 1989, 1993). Figure 3.4 shows the results for the few specimens for which this is available, again keyed to a chimpanzee standard value of 1 (Foley 1996).

A third indicator of behavioral capabilities and their underlying cognitive basis is that of technology. It is well established that stone tool assemblages change over time in ways that can be interpreted in terms of increased complexity, longer operational sequences, and greater variation in end product (Schick and Toth 1993). Archaeologists have classified these stone tools in a number of ways, grouping them according to mode of production and the typology of the finished tools. We have argued elsewhere that there is a reasonably good match between stone tool technology in terms of mode of production and hominin taxa and their proposed phylogenetic relationships (figure 3.5) (Foley 1987; Foley and Lahr 1997). Early *Homo*, early *Homo ergaster*, and *Homo erectus* are all characterized by mode 1 (simple pebble tools and Oldowan industries); later *H. ergaster* developed mode 2 (bifaces, hand axes, and the Acheulean) technology, which shows a more symmetrical, complex, and patterned process of manufacture. This is also found in *Homo heidelbergensis*. Mode 3 technologies (prepared core, the Levallois, and Middle Paleolithic/Middle Stone age tools, with an emphasis on points and flake-based tools) are shared by both the Neanderthal and

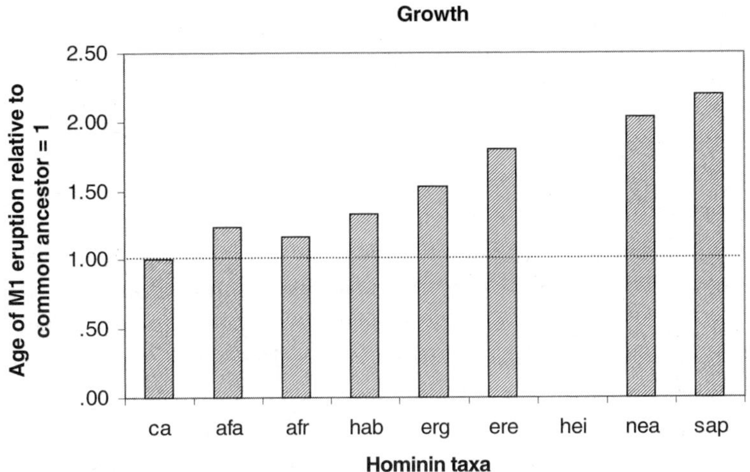

Figure 3.4

Measures of the relative rate of maturation of extinct hominin taxa, expressed as a ratio to the hypothesized common ancestor with *Pan*. Data based on rate of M1 eruption. See Foley (1996) for detail. Key: See figure 3.3.

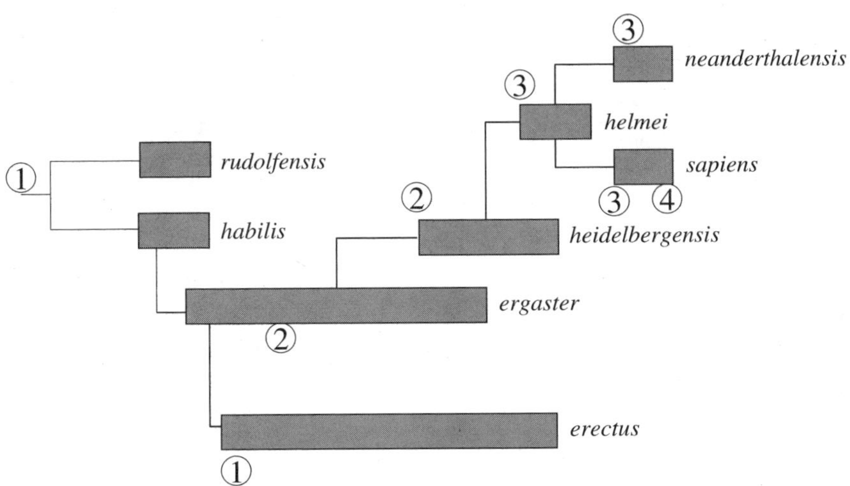

Figure 3.5

Phylogeny of *Homo* showing distribution of technological modes. Key to modes: 1, Oldowan and pebble tool industries; 2, bifaces and Acheulean; 3, prepared core; 4, blade-dominated technology.

sapiens lineages, and occur in their proposed common ancestor (*H. helmei*). As discussed earlier, full mode 4 technologies are found only sporadically in early *Homo sapiens* and make a full appearance in the European Upper Paleolithic.

These traditions display two other observable traits: the length of time they persist and the number of tool types they produce. The pattern in these is shown in figure 3.6. These figures are approximate and are absolute figures (Foley 1996). The nature of chimpanzee technology and an absence of any information concerning the longevity of their traditions makes it harder to calculate the figures in the same manner as those for brain size or growth rate, although the absence of any known stone tool manufacture among chimpanzees and the lack of certainty for such manufacture by australopithecines, make *Homo* in effect the baseline.

If we assume that these observable features reflect to some degree the cognitive capacities of the hominins, then we can perhaps make some inferences about the preconditions upon which the full cultural capacities are based and the way in which they were acquired. Two general points should be made initially. The first is that in line with the earlier discussion, there is no evidence for a single quantum leap in hominin abilities. Each new taxon or archaeological unit shows an incremental increase over others, but these appear at different points in time, often with prolonged periods of stability in between (see later discussion). It is not the case that there is a smooth and general change, bearing in mind that the data shown in figures 3.3, 3.4, and 3.6 are not plotted against time (but see figure 3.7). The second point is that although there is some correlation among the different traits (see Foley 1996), they also show different patterns according to which trait is measured. Life history strategies appear to change relatively evenly across the taxa, and it can be argued that the underlying biological relationships affecting body size, brain size, rate of maturation, and presumably changes in parental investment are under constant selective pressure throughout the period of hominin evolution. This would perhaps argue for this being a major precondition on which other traits are superimposed. The EQ, on the other hand, shows a pattern in which relative brain size increases most dramatically after *Homo heidelbergensis,* and this may be perhaps relative to the achievement of virtually full modern human life history characteristics by this time (Foley and Lee 1991).

Changes in technological modes are, by definition in the way they are measured, more prone to be major ones, and in this case the most significant change is seen in the longevity of the Middle Stone Age/Middle Paleolithic (the European Acheulean pattern is probably an artifact of colonization events). A similar but less marked pattern can be seen in the technological complexity.

The answer to the question of what pattern of behavioral and cognitive change might reflect cultural predispositions is therefore that there appear to be marked differences between *Homo* and the australopithecines in many characteristics, as has been noted before, and that within *Homo* there is a relatively gradual shift in life history

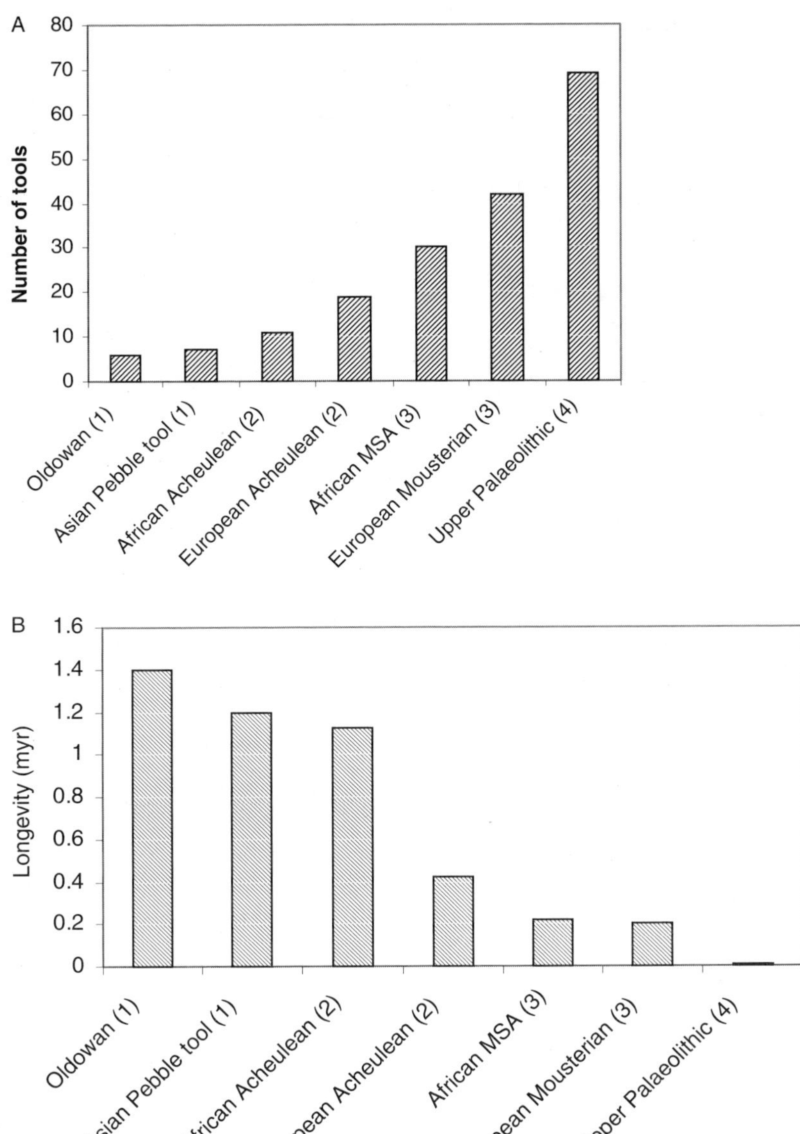

Figure 3.6

(A) Measures of the longevity (duration from first to last appearance) and (B) complexity (number of stone tool types) for major archaeological units. See Foley (1996) for details.

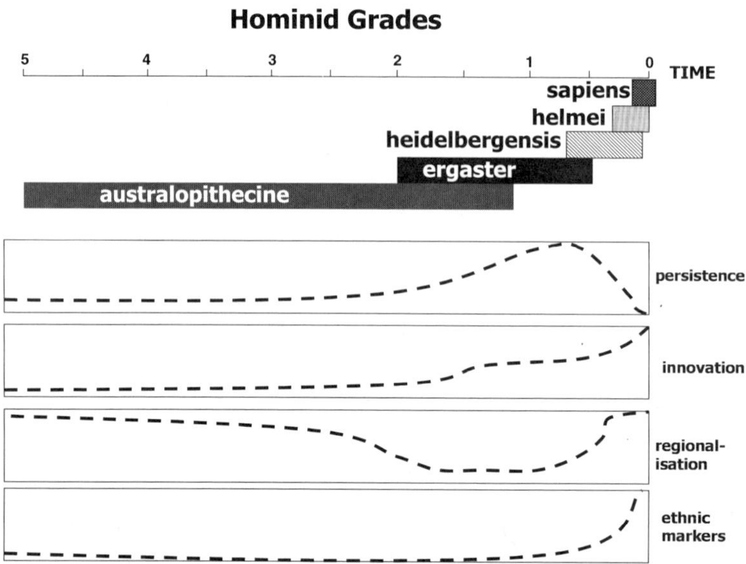

Figure 3.7
Temporal estimates of the evolution of some characteristics of the archaeological record and inferred hominin grades. The dotted lines show an estimate of similarity to the modern human condition. These patterns show that although change is spread across much of the Pleistocene and Pliocene, it is not a continuous process.

strategy, perhaps reflecting the continual selective pressures related to the costs of brain growth. Finally, in both brain and technological measures, the period between the emergence of *Homo heidelbergensis* and the full appearance of both Neanderthals and modern humans (i.e., somewhere between 400,000 and 200,000 years ago) is critical for the major changes. However, it should be noted that across this time period the changes observed are still occurring on a relatively slow, i.e., evolutionary scale, rather than in the manner that one might expect for cultural change in the form we associate with fully modern humans.

What Was the "Cultural Status" of Different Hominin Taxa?

On the basis of this general pattern of evidence, we might now consider briefly what might be the actual cultural status of the various hominin taxa that have been recognized. Given the nature of the evidence, this must be speculative at best, but it may perhaps provide insights into which elements of cultural cognition came into play at various points in the past. The evidence observed does not suggest that there is a single modern package of cultural capacities that appears full blown, and so it is worth examining how the "package" may have been put together in extinct hominins.

Table 3.2 summarizes the available information and major inferences. The baseline for comparative purposes, despite the well-known limitations of the underlying assumptions, is the chimpanzee. Although there is no means of knowing what traits in living chimpanzees may have evolved at a later date, and so were not present in the common ancestor, such an approach is perhaps defensible in terms of the fact that many of the characteristics are found in both chimpanzees and humans, and so on the grounds of parsimony, may be assumed to be plesiomorphic. It is reasonable to assume, therefore, that they were also present among australopithecines, and there is no substantial evidence to suggest that this latter group was in any way significantly different from chimpanzees in these cognitive and social abilities.

Homo ergaster shows two major trends away from this baseline: first in the complexity of the tools made (mode 2), and second in terms of delayed growth. However, what is also striking about these tools is their persistence, standardization, and lack of innovation over long periods and across vast distances. From this we may perhaps infer that *Homo ergaster* was capable of longer chains of planning and had the ability to perceive different forms within others and to organize the acquisition and production of materials. From the life history changes we might also infer that there was greater maternal investment and longer mother-infant bonds (Foley 1995), and these might have prompted greater levels of emotional affiliation. If, as has been argued, the meat-eating evidence also implies more provisioning by males, then these stronger affiliations may also have been present between adults. If this is the case, then emotional attachments may be an early element of human cognition and behavior. However, this occurred in the context of what appears to have been strongly stereotypical behavior and an emphasis on the ability to imitate others accurately, rather than to explore variable outcomes. Although no doubt still unproven, there is also a case that *H. ergaster* did not possess the ability to use language (McLarnon 1996; Kay et al. 1998).

Homo heidelbergensis shows few behaviors that could be said to depart strongly from those seen in *H. ergaster*. The major behavioral change appears to have been greater control of fire, if not the first active use of it, allowing the occupation of higher and colder latitudes (Klein 1999). The cognitive implications of fire have not been fully explored or investigated, but anthropologists have long recognized the importance that fire can play in maintaining social relationships. In terms of technology, it can also be argued that their handaxes have greater standardization and symmetry. The other key element of *Homo heidelbergensis* is that it may have evolved life history parameters very similar to those of modern humans (Foley and Lee 1991), thus embedding still further the emotional alliances that it has been argued occur in *Homo ergaster*.

Homo helmei, or less contentiously, the later African larger-brained archaic forms, do show several innovations, ones that are found in Neanderthals and early modern humans (Foley and Lahr 1997; Lahr and Foley 1998). There is a strong probability that

Table 3.2

Hypothesized "Cultural" Properties of Hominin Taxa and Some Empirical Bases

	Common ancestor and australopithecines	H. ergaster	H. heidelbergensis	H. helmei H. neanderthalensis	H. sapiens
"Cultural" inference		Greater planning depth Imitation Limited innovation Emotional affiliation? Theory of mind?	Parental care?	Greater planning depth Ethnic affiliation Symbolic thought Language	Strong ethnicity Extensive symbolism
Observation	Communities Fission-fusion Male kin bonding Territoriality Political alliances Basic tool making Ephemeral traditions	Significant meat eating Delayed growth Complex tool making Persistent traditions	Fire? Modern life history?	Flexible technology Regionalization	Local networks Cultural replacements Rapid change Material diffusion

they were capable of language in some form (Aiello and Dunbar 1993; McLarnon 1996; Kay et al. 1998), and this would certainly imply some degree of symbolic thought. Furthermore, their stone technology was more complex and, as has been shown (Deacon and Shuurman 1992; McBrearty and Brooks 2000), a number of modern features are found in the African populations—use of bone tools, pigments, etc. The greater regionalization of technologies and the reduced longevity of traditions may also be indicative of greater populational differences being determined culturally and the development of forms that are akin to modern ethnicity and cultural exclusivity (Mellars 1991).

It has to be admitted that there is little in the archaeology of very early modern humans that indicates capacities that are not present in the Neanderthals or their last common ancestor. Among the possibilities that have been suggested for the African Middle Stone Age are the use of bone tools; pigments for body decoration; and a much greater flexibility in the use of stone tool technologies, showing still greater regionalization. There is also a tendency for traditions to last only a few thousand years, rather than as before, tens or even hundreds of thousands of years; this can certainly be seen strongly in the populations of the later Pleistocene and early Holocene (Deacon 1992; Klein 1992, 1995; McBrearty and Brooks 2000).

In summary, therefore, the fossil and archaeological evidence points to the following: (1) Australopithecines were probably at the same cultural and cognitive grade as living chimpanzees. (2) Through the period of *Homo ergaster* and *H. heidelbergensis* there was a driving pattern of delayed maturation, with implications for the emotional basis of social relationships. (3) At two stages, with the development of mode 2 and mode 3 technologies, there is evidence for first strong imitation and stereotypical behaviors within a pattern associated with longer chains of planning, and second, greater flexibility and greater regionalization of populations on the basis of their behaviors. (4) Language and communication, which are presumably associated with greater capacity for symbolic thought, also occur at some point in the period about a quarter of a million years ago.

The final point to be made, however, is that although these basic biological capacities were in place at these considerable ages, it was only very much later that they led to the explosive pattern of cultural evolution seen in recent humans. The most important conclusion to be drawn is that species-specific biology alone is not enough. It must occur in specific ecological and demographic contexts, perhaps associated with larger groups, higher population densities, and greater intergroup competition, for the material expression of cultural potential and complexity to occur.

What Was "Happening" as Cultural Properties Evolved?

If the capacities for cultural behavior evolved piecemeal during the Pleistocene, then there are two major questions that need to be considered. The first is, what were the selective conditions that prompted it? Why did a pattern of greatly enhanced behavioral flexibility based on cognitive depth occur in this particular lineage (and no other)? The second question is, exactly what is it that is being selected for in the process? While these two questions are at present unanswerable using the available evidence, nonetheless some suggestions about the direction in which answers may be sought can be tentatively suggested.

The recurrent theme in the observed pattern of the evolution of *Homo* appears to be a delayed process of maturation, and this has been linked to the idea that it allows greater neurobiological plasticity and therefore a greater scope for learning during the course of development. The question this prompts is, learning for what? One answer could be social complexity, and that this itself is driving life history evolution (Dunbar 1992). However, social complexity is not in itself something that is selectively advantageous. A better answer might be to consider the correlates of delayed maturation above and beyond brain size and behavioral plasticity, correlates that can be more closely linked to reproductive success. The prime possibility in this respect is longevity (Austad and Fischer 1992; Allman et al. 1993).

Increased longevity will increase the adult life-span even more than it increases the period of development, and this will also mean an increase in mating opportunities. Any extension of life-span is likely to have a positive effect on reproductive effort. This is particularly the case for males, where the last few years of reproduction may be the most productive (there may be different constraints operating on female reproductive success; see Turke 1997, and Hawkes et al. 1998). If this is the case, then it could be argued that the selective pressures leading to delayed maturation are in fact a by-product of selection for adult longevity, and that the developmental advantages that arise and allow greater behavioral flexibility as well as overall increased cognitive abilities are secondary but bring advantages to a long-lived species; i.e., advantages gained in living in increasingly complex social groups with multiple generations. There is some support for this model; Kaplan et al. (2000) have shown that among contemporary hunter-gatherers, male longevity is strongly related to both hunting success and number of surviving offspring. Furthermore, brain size, the measurable sign of the cognitive difference between humans and other species, is strongly related to longevity once body size has been excluded (figure 3.8).

A further component of this argument relates to the social context in which such selection is taking place. It has been shown that female kin-bonded groups are absent from the hominoids, but are widespread among cercopithecoids (Foley and Lee 1989; Rendall and Di Fiore 1996). This has led some to argue that male kin bonding, which is found in the human sister clade, *Pan*, is a key part of human social evolution (Wrangham 1979; Foley and Lee 1989). In male kin-bonded groups, single males do not monopolize reproduction among a group of females, but they probably have reproductive access for a longer period of time. It may well be that this is the context in which

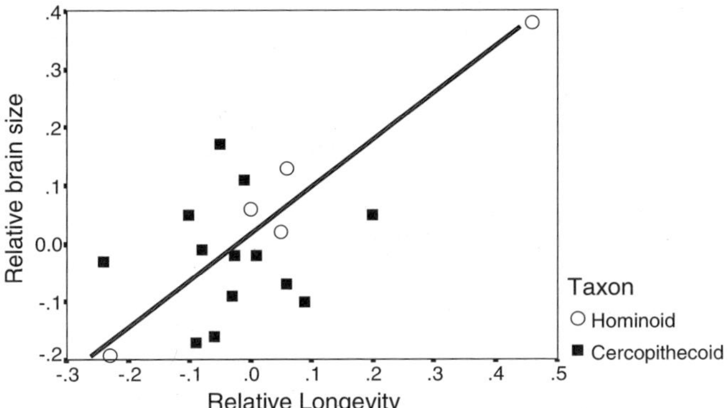

Figure 3.8
Relative longevity of catarrhines plotted against relative brain size. For hominoids there is a strong positive relationship, but no such relationship exists for cercopithecoids.

male longevity, and thus also changes in life history, will occur, and why it is among a hominoid primate that we see the trend observable in the hominin lineage. It is interesting to note in this context that among the cercopithecoids there is no relationship between brain size and longevity when effects of body size are removed (figure 3.8).

If the interaction of selection for longevity and changes in overall life history parameters is the key to the selective pressures leading to a "cultural primate," then one further point to establish is the ecological context. It has been shown that large brains are generally constrained by the energy costs associated with their growth. It follows that delayed maturation and larger brains impose greater metabolic costs on mothers, and these must be supplied by a reliable and high-quality food resource. In this context it is noticeable that one of the factors that seems to distinguish *Homo* from australopithecines is evidence for increased levels of meat eating (Bunn and Kroll 1986), and furthermore, that at the time of the major expansion in brain size there is a shift to better hunting projectiles and a wider range of prey sources (Stiner et al. 1999).

If selection for increased longevity, especially among males, lies at the heart of the evolutionary trends seen in the genus *Homo*, it remains to be seen what exactly might be selected for in terms of cultural propensities. There has been considerable debate about this, focusing in recent years on ideas relating to the social mind, a theory of mind, and specialized modules (Byrne and Whiten 1986; Barkow et al. 1992; Dunbar 1992). It is likely that all of these have a role to play and that as yet it is not possible to untangle the various elements. However, it is probably the case that these different elements are to some extent hierarchically organized in terms of the ways in which they are responses to the selective pressures and environments in which hominins were living. In particular, there is likely to have been parallel evolution of both general

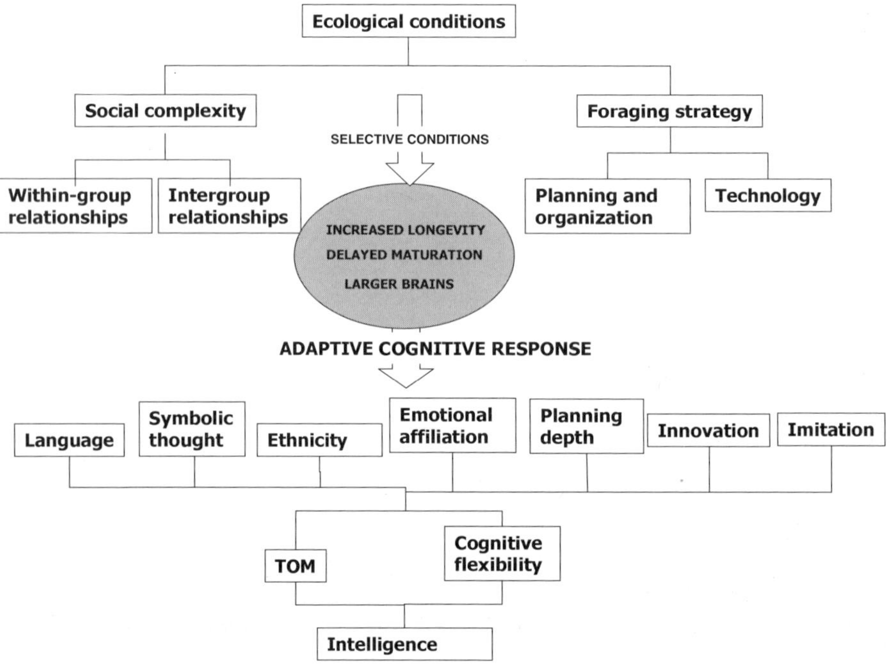

Figure 3.9

A model of cultural evolution. In this model, selective pressures operate through selection for male longevity and other life history parameters occurring in particular socioecological conditions. Culture may be considered to be the totality of the cognitive response and itself a focus of parallel selection. See text for discussion.

characteristics of intelligence and flexibility and specialized ones. Given the fact that hominin evolution shows changes across a range of activities, it is likely that all of these will have changed in response to the way in which longevity, or some other factor, was altering the hominin cognitive environment. The effects will have been across the range—from emotional affiliation between individuals to the balance between innovation and imitation, in social, technological, and economic realms (figure 3.9).

Conclusions

In this chapter I have addressed the issue of the emergence of culture from the perspective of paleobiology and archaeology; that is, the direct evidence for the preconditions upon which modern cognition and cultural behavior are based. It should be stressed that this is of considerable importance as an approach and also tantalizingly imprecise. It is important because the extinct hominins are far closer to humans than the living primates, and so by definition undermine arguments that there is an unbridgeable gap between humans and other animals. What limits this approach, however, is the sparse and incomplete data available, which means that many important elements are invisible. These limitations should be borne in mind.

However, a number of moderately robust conclusions can be drawn. The first is that human cultural and cognitive evolution was not a single revolutionary event associated with the origins of modern humans. There is little doubt that the fossil hominins, had they survived, would have shown characteristics intermediate between those of humans and chimpanzees. However, these archaic hominins were not all the same; some, such as the Neanderthals, were clearly similar to modern humans in many ways, whereas others, specifically the australopithecines, may have been little different from chimpanzees. This does not mean that it is simply a question of continuous gradual change, because these shifts may have been concentrated in a few critical periods and events, particularly in the last half million years. Rather than one revolutionary change, there may have been a set of cumulative and individually significant cognitive and behavioral events.

A second conclusion is that changes in patterns of growth and development seem to show an overall trend. While the samples on which this observation is based are limited, nonetheless this seems to be a critical and underlying change. I have argued here that this is most probably related to selection for increased longevity, especially among males, and that it is this that has dragged along prolonged development, which may have enhanced the potential for learning and neural plasticity, as well as the energetics of larger brain sizes.

The third conclusion is that in examining the evolution of culture, there are two components that have to be considered. On the one hand, there are the selective conditions that shape behaviors leading to survivorship and reproduction, and these will be context specific; they are likely to be strongly interlinked social and ecological factors. On the other hand, there is the selection for the cognitive capabilities that may be considered at the response to those conditions. Cultural evolution is not merely "cultural" or "cognitive," but is rooted in biology, ecology, and behavior.

References

Aiello, L., and Dunbar, R. I. M. 1993. Neocortex size, group size and the origin of language in the hominids. *Current Anthropology* 34: 184–193.

Allman, J., McLaughlin, T., and Hakeem, A. 1993. Brain structures and life-span in primate species. *Proceedings of the National Academy of Sciences of the USA* 90: 3559–3563.

Austad, S. N., and Fischer, K. E. 1992. Primate longevity: Its place in the mammalian scheme. *American Journal of Primatology* 28: 251–261.

Barkow, L., Cosmides, L., and Tooby, J. (eds.) 1992. *The adapted mind*. New York: Oxford University Press.

Beynon, A. D., and Dean, M. C. 1988. Distinct dental development patterns in early fossil hominids. *Nature* 335: 509–514.

Bromage, T. G., and Dean, M. C. 1985. Re-evaluation of the age at death of Plio-Pleistocene fossil hominids. *Nature* 317: 525–528.

Bunn, H. T., and Kroll, E. M. 1986. Systematic butchery by Plio/Pleistocene hominids at Olduvai Gorge, Tanzania. *Current Anthropology* 27: 431–452.

Byrne, R., and Whiten, A. (eds.) 1986. *Machiavellian intelligence.* Oxford: Clarendon Press.

Cavalli-Sforza, L., and Feldman, M. 1981. *Cultural transmission and evolution.* Princeton, N.J.: Princeton University Press.

Clutton-Brock, T. H. 1991. *The evolution of parental care.* Princeton, N.J.: Princeton University Press.

Davidson, I., and Noble, W. 1991. The evolutionary emergence of modern human behavior: Language and its archaeology. *Man* 26: 223–254.

Deacon, H. J. 1992. Southern Africa and modern human origins. *Philosophical Transactions of the Royal Society of London* B337: 177–183.

Deacon, H. J., and Shuurman, R. 1992. The origins of modern people: The evidence from Klasies River. In G. Brauer and F. H. Smith (eds.), *Continuity or replacement? Controversies in Homo sapiens evolution* (pp. 121–129). Rotterdam: Balkema.

Dean, M. C., Stringer, C. B., and Bromage, T. G. 1986. Age at death of the Neanderthal child from Devils Tower, Gibraltar and the implications for studies of general growth and development in Neanderthals. *American Journal of Physical Anthropology* 70(3): 301–309.

Dunbar, R. I. M. 1992. Neocortex size as a constraint on group size in primates. *Journal of Human Evolution* 22: 469–493.

Foley, R. A. 1987. Hominid species and stone tool assemblages: How are they related? *Antiquity* 61: 380–392.

Foley, R. A. 1991. How useful is the culture concept in early human studies? In R. A. Foley (ed.), *The origins of human behaviour* (pp. 25–38). London: Unwin Heinmann.

Foley, R. A. 1995. The evolution and adaptive significance of human maternal behaviour. In C. Pryce, R. D. Martin, and D. Skuse (eds.), *Human and non-human primate mothers: An integrated approach* (pp. 27–36). Zürich: Karger.

Foley, R. A. 1996. Measuring the cognition of extinct hominids. In P. Mellars and K. Gibson (eds.), *Modelling the early human mind* (pp. 57–66). Cambridge: MacDonald Institute.

Foley, R. A. 1998. Pattern and process in hominid evolution. In J. Bintliff (ed.), *Structure and contingency: Evolutionary processes in life and human society* (pp. 31–42). London: Leicester University Press.

Foley, R. A. 2001. Evolutionary perspectives on the origins of human social institutions. *Proceedings of the British Academy* 110: 171–195.

Foley, R. A., and Lahr, M. M. 1997. Mode 3 technologies and the evolution of modern humans. *Cambridge Archaeological Journal* 7: 3–36.

Foley, R. A., and Lee, P. C. 1989. Finite social space, evolutionary pathways and reconstructing hominid behavior. *Science* 243: 901–906.

Foley, R. A., and Lee, P. C. 1991. Ecology and energetics of encephalization in hominid evolution. *Philosophical Transactions of the Royal Society of London* B334: 223–232.

Goldstein, D. B., Linares, A. R., Cavalli-Sforza, L. L., and Feldman, M. W. 1995. Genetic absolute dating based on microsatellites and the origin of modern humans. *Proceedings of the National Academy of Sciences of the USA* 92(15): 6723–6727.

Hammer, M. F. 1995. A recent common ancestry for human Y chromosomes. *Nature* 378: 376–378.

Harpending, H., Sherry, S. T., Rogers, A. R., and Stoneking, M. 1993. The genetic structure of ancient human populations. *Current Anthropology* 34: 483–496.

Harvey, P. H., and Clutton-Brock, T. H. 1985. Life history variation in primates. *Evolution* 39: 559–581.

Harvey, P. H., and Pagel, M. D. 1988. The allometric approach to species differences in brain size. *Human Evolution* 3: 461–472.

Harvey, P., and Pagel, M. 1992. *The comparative method in evolutionary biology*. Oxford: Oxford University Press.

Harvey, P. H., Martin, R. D., and Clutton-Brock, T. H. 1987. Life histories in comparative perspective. In B. B. Smuts, D. L. Cheney, R. M. Seyfarth, R. W. Wrangham, and T. T. Struhsaker (eds.), *Primate societies* (pp. 181–196). Chicago: University of Chicago Press.

Harvey, P. H., Promislow, D. E. L., and Read, A. F. 1989. *Causes and correlates of life history differences among mammals*. Oxford: Blackwell.

Hawkes, K., O'Connell, J. F., Jones, N. G. B., Alvarez, H., and Charnov, E. L. 1998. Grandmothering, menopause, and the evolution of human life histories. *Proceedings of the National Academy of Sciences of the USA* 95(3): 1336–1339.

Henshilwood, C., and Sealy, J. 1997. Bone artefacts from the Middle Stone Age at Blombos Cave, southern cape, South Africa. *Current Anthropology* 38(5): 890–895.

Henshilwood, C., d'Errico, F., et al. 2004. Middle Stone Age shell beads from South Africa. *Science* 384: 404.

Holloway, R. L. 1983. Cerebral brain endocast pattern of *Australopithecus-afarensis* hominid. *Nature* 303(5916): 420–422.

Hublin, J. J., Spoor, F., Braun, M., Zonneveld, F., and Condemi, S. 1996. A late Neanderthal associated with Upper Paleolithic artefacts. *Nature* 381(6579): 224–226.

Ingman, M., Kaessmann, H., Paabo, S., and Gyllensten, U. 2000. Mitochondrial genome variation and the origin of modern humans. *Nature* 408(6813): 708–713.

Jerison, H. J. 1973. *Evolution of the brain and intelligence.* New York: Academic Press.

Kaplan, H., Hill, K., Lancaster, J., and Hurtado, A. M. 2000. A theory of human life history evolution: Diet, intelligence, and longevity. *Evolutionary Anthropology* 9(4): 156–185.

Kay, R. F., Cartmill, M., and Balow, M. 1998. The hypoglossal canal and the origin of human vocal behavior. *Proceedings of the National Academy of Sciences of the USA* 95(9): 5417–5419.

Klein, R. G. 1992. The archaeology of modern human origins. *Evolutionary Anthropology* 1: 5–14.

Klein, R. G. 1995. Anatomy, behavior and modern human origins. *Journal of World Prehistory* 9: 167–198.

Klein, R. G. 1999. *The Human Career* (2nd ed.). Chicago: University of Chicago Press.

Klein, R. G. 2000. Archaeology and the evolution of human behavior. *Evolutionary Anthropology* 9(1): 17–36.

Knight, C., Powers, C., and Watts, I. 1995. The human symbolic revolution: A Darwinian account. *Cambridge Journal of Archaeology* 3: 75–114.

Lahr, M. M., and Foley, R. A. 1994. Multiple dispersals and modern human origins. *Evolutionary Anthropology* 3: 48–60.

Lahr, M. M., and Foley, R. A. 1998. Towards a theory of modern human origins: Geography, demography, and diversity in recent human evolution. *Yearbook of Physical Anthropology* 41: 137–176.

Martin, R. D. 1981. Relative brain size and basal metabolic rates in terrestrial vertebrates. *Nature* 293: 57–60.

Martin, R. D. 1983. *Human brain evolution in an ecological context.* New York: American Museum of Natural History.

McBrearty, S., and Brooks, A. S. 2000. The revolution that wasn't: A new interpretation of the origin of modern human behavior. *Journal of Human Evolution* 39(5): 453–563.

McBurney, C. B. M. 1967. *The Haua Fteah (Cyrenaica) and the Stone Age of the South East Mediterranean.* Cambridge: Cambridge University Press.

McGrew, W. C. 1992. *Chimpanzee material culture: Implications for human evolution.* Cambridge: Cambridge University Press.

McLarnon, A. 1996. The evolution of the spinal cord in primates: Evidence from the foramen magnum and the vertebral canal. *Journal of Human Evolution* 30: 121–138.

Mellars, P. 1991. Cognitive changes and the emergence of modern humans in Europe. *Cambridge Archaeological Journal* 1: 63–76.

Mithen, S. 1996. *The prehistory of the mind.* London: Thames and Hudson.

Rendall, D., and Di Fiore, A. 1996. The road less traveled: Phylogenetic perspectives in primatology. *Evolutionary Anthropology* 4(2): 43–52.

Schick, K., and Toth, N. 1993. *Making silent stones speak.* New York: Simon & Schuster.

Smith, B. H. 1989. Dental development as a measure of life history in primates. *Evolution* 43: 683–688.

Smith, B. H. 1993. The physiological age of WT15000. In A. C. Walker and R. E. Leakey (eds.), *The Nariokotome skeleton* (pp. 195–220). Cambridge, Mass: Harvard University Press.

Stiner, M. C., Munro, N. D., Surovell, T. A., Tchernov, E., and Bar-Yosef, O. 1999. Paleolithic population growth pulses evidenced by small animal exploitation. *Science* 283(5399): 190–194.

Stringer, C. B. 1989. The origin of modern humans: A comparison of the European and non-European evidence. In P. Mellars and C. B. Stringer (eds.), *The human revolution* (pp. 232–244). Edinburgh: Edinburgh University Press.

Stringer, C. B., and Andrews, P. 1988. Genetic and fossil evidence for the origin of modern humans. *Science* 239: 1263–1268.

Tattersall, I. 2000. Once we were not alone. *Scientific American* 282(1): 56–62.

Tomasello, M. 2000. *The cultural origins of human cognition.* Cambridge, Mass: Harvard University Press.

Turke, P. W. 1997. Hypothesis: Menopause discourages infanticide and encourages continued investment by agnates. *Evolution and Human Behavior* 18(1): 3–13.

Valladas, H., Joron, J. L., Valladas, G., Arensburg, B., Bar-Yosef, O., Belfer-Cohen, A., Goldberg, P., Laville, H., Meignen, L., Rak, Y., Tchernov, E., Tillier, A.-M., and Vandermeersch, B. 1987. Thermoluminescence dates for the Neanderthal burial site at Kebara in Israel. *Nature* 330: 159–160.

Vigilant, L., Stoneking, M., Harpending, H., Hawkes, K., and Wilson, A. 1991. African populations and the evolution of human mitochondrial DNA. *Science* 253: 1503–1507.

Whiten, A., Goodall, J., McGrew, W. C., Nishida, T., Reynolds, V., Sugiyama, Y., Tutin, C. E. G. Wrangham, R. W., and Boesch, C. 1999. Cultures in chimpanzees. *Nature* 399(6737): 682–685.

Wrangham, R. W. 1979. On the evolution of ape social systems. *Social Science Information* 18: 335–368.

4 Interactions of Culture and Natural Selection among Pleistocene Hunters

Christopher Boehm

Interactions between natural selection and cultural processes have been studied for several decades, with increasing emphasis on specific case studies. The modern pioneer was Donald T. Campbell (1965b, 1972, 1975), who likened the self-organizing side of cultural processes to blind variation and selective retention in biological processes. Subsequently Boyd and Richerson (1982, 1985) defined these self-organizing processes in terms of conformist transmission mechanics (see also Cavalli-Sforza and Feldman 1981; Richerson and Boyd 2005; Boyd and Richerson 1991), while Boehm (1978, 1991, 1996) has emphasized the intentional side of cultural selection by focusing on individual and group decisions. Finally, Durham (1976, 1982, 1991) has broadly integrated the field of human gene–culture coevolution, using case studies that take account of both self-organizing and intentionally guided processes.

Here I offer some hypotheses about one major phase of this interactive process, a phase that involved the emergence of the first human moral communities. The hypothesis is that over time, a moral approach to keeping social order began to affect the very process of natural selection, and that this changed human nature in ways that profoundly influenced morality as we know it today. To understand this sequence of events, it is necessary to start at the beginning, with an ancestral ape.

Cladistics

A four-species behavioral reconstruction model will be set up by which any behavior that is shared by all four African-based hominoids (humans, chimpanzees, gorillas, and bonobos) can be posited as being present in the earlier hominoid that was ancestral to these species. Wrangham (1987) developed an experimental behavioral model of this ancestor, and the modeling was highly conservative because it is so unlikely that convergent evolution could cause all four species in a closely related clade to arrive at a new behavior that was absent ancestrally. The coevolutionary analysis in this chapter will be anchored by a rather complete description of Wrangham's four-species common ancestor's behavioral repertoire.

I will borrow from a prior analysis (Boehm 1999b), in which this type of reconstruction was used to identify preadaptations useful to the evolutionary emergence of political egalitarianism. Preadaptation involves putting old traits to new uses as environments change, so we are interested not only in how the common ancestor was likely to have behaved in its actual environments but also in any behavioral potential that could have been stimulated by new environments. Because of this interest in preadaptive potential, in focusing on ancestral traits I will consider not only how the four descendent species behave in nature, but how three of them, the African great apes, behave in novel environments, notably in captivity.

Ancestral Use of Resources

It is agreed that the common ancestor was a quadrupedal arboreal ape whose main diet was fruit or leaves (see Foley 1997; Tattersall 1993). However, we are guessing at many other aspects of the subsistence pattern because the four descendent hominoids vary considerably. Human hunter-gatherers depend substantially on flesh in their diet (Kelly 1995); chimpanzees do so to a moderate but highly significant degree (Stanford 1999); and bonobos do a certain amount of hunting (Kano 1992; Wrangham and Peterson 1996); by contrast, gorillas at best pick up the occasional insect (Fossey 1983). Thus, whereas the more recent-*Pan*-Human ancestor (see Boehm 2000a; Wrangham and Peterson 1996) was likely to have been a hunter (Stanford 1998, 1999) the earlier common ancestor must be left as indeterminate as to whether meat was included in its diet. We simply cannot tell whether gorillas lost a compelling interest in animal fat and protein or whether the other three species developed it after gorillas split off.

Common ancestral foraging groups were closed, with definite social boundaries (Wrangham 1987), and they probably were smallish, ranging from about a dozen to a hundred or so. Human hunters, chimpanzees, and bonobos live in fission-fusion groups, with continual formation of ad hoc subgroups whose size is determined by foraging conditions, whereas gorillas essentially stay together as a single harem unit (see Boehm 1999b). Thus, the common ancestor lived in smallish closed groups that moved around, foraging for food, while the *Pan*-human ancestor can be further specified as having a fission-fusion social structure.

Territorially, the picture is less clear. Human foragers often defend natural resources (see Ember 1978; Dyson-Hudson and Smith 1978; Cashdan 1983), as do chimpanzees (Goodall 1979; Nishida 1979). Bonobos show some tendencies in this direction (Kano 1992; Wrangham and Peterson 1996), but because gorillas are not observed to compete for natural resources, the common ancestor must be left as a question mark, territorially speaking.

A Political Portrait of the Common Ancestor

When Wrangham (1987) developed the four-species common ancestral behavioral model, relatively little was known about bonobos or lowland gorillas, and he only surveyed accounts of natural behavior, leaving captives aside. Although he identified several basic patterns of social behavior that included closed groups and stalking and killing of conspecifics by males, coalition behaviors were taken as being dubious or absent in females and males, and one fundamental type of social behavior was not evaluated at all. This was the social dominance hierarchy (see Tiger and Fox 1971), with the status rivalry that underlies it.

In retrospect, this descriptive omission is understandable. Even though chimpanzees, gorillas, and bonobos clearly live in hierarchies, in 1987 the case of humans was rather ambiguous. Many scholars considered hunter-gatherers, the appropriate humans for comparison, to be naturally egalitarian, as opposed to innately hierarchical. This was the case even though a few scholars such as Fried (1967) and Service (1962, 1975) had noted that hierarchy was more muted than absent (see also Flanagan 1989); while others had been implying (Lee 1979; Woodburn 1979) or asserting (Boehm 1982b, 1984; Woodburn 1982) that egalitarianism involved the use of coercive subordinate force to keep down would-be individual dominators.

In 1993, building on other relevant work (e.g., Cashdan 1980; Fried 1967, Gardner 1991; Service 1962; Woodburn 1982), I advanced a formal hypothesis: Egalitarian hunter-gatherers and the egalitarian tribesmen who followed them were just as hierarchically inclined as humans living in chiefdoms or states, but because the subordinates were so firmly (and dominantly) in control, this gave a false impression of nonhierarchy. For a long time, this impression had been reinforced by a scholarly propensity to find human precursors who were "just naturally equal," and there was initial resistance to this interpretation (e.g., commentaries in Boehm 1993; Erdal and Whiten 1994, 1996; see also Wiessner 1998). However, the general hypothesis now appears to be widely accepted (e.g., Knauft 1994; Wiessner 1996; commentaries on Boehm 2000a). I will therefore amplify the scope of Wrangham's appraisal by assuming humans to be just as involved with status rivalry and just as hierarchically inclined as the other three african apes (Boehm 1999b). This makes the common ancestor hierarchical as well, and this feature is critical to the arguments that follow because dominance and punishment are intrinsic to social control.

The common ancestral political model I am building here is based, not only on a new interpretation of hunter-gatherer politics as being heavily involved with their own brand of hierarchical behavior (Boehm 1993), but on more recent findings from primatological studies in the wild, and on abundant and fascinating data on behavior in captivity. Furthermore, whereas Wrangham's model was basically "social," the model I present here is specifically sociopolitical. I am interested in the uses of individual and collective power in group life because they played a critical role in the evolutionary development of morality.

The common ancestor engaged in political coalition behavior (see Boehm 1999b), which is prominent in humans and wild and captive chimpanzees (de Waal 1982; Goodall 1986), is noteworthy in wild and captive bonobos, and definitely is seen with gorillas. Bonobo females form coalitions that enable them to compete with males in certain behavioral contexts such as feeding (Kano 1992; Stanford 1999), while bonobo males tend to hang together politically when two communities meet (see Wrangham and Peterson 1996). In multimale harems, male gorillas form coalitions to defend the harem, whereas in captivity, females have been seen to unite and drive away a new silverback when they already had a younger male who was leading them (de Waal 1982). Because all four extant species form male or female coalitions that are larger than dyadic partnerships, this behavioral potential was present ancestrally and therefore was present in the human line from the beginning.

Conflict Resolution as a Key Preadaptation

Another behavior present in all four hominoids is pacifying interventions in conflicts (Boehm 1999b). Humans in bands at least try to intervene triadically, although they are not very successful (Furer-Haimendorf 1967); chimpanzee males and females of higher rank intervene in a pacifying mode (Boehm 1994; de Waal 1982), and the same is true of silverback gorillas regulating their harem females (Fossey 1983). Bonobos are less well studied in this respect, but similar patterns have been observed.

This innate aversiveness to conflict within the group, along with the active tendency to manipulate protagonists in the direction of pacification, was an important preadaptation for moral behavior. Indeed, people in hunting bands dislike conflict, and behaviors they single out for negative sanctioning are mostly ones that lead to conflict (Boehm 1982b, 1999b). Furthermore, de Waal's (1989) work on peacemaking among primates (see also Cords 1997; de Waal 1996; Aureli and de Waal 2000) suggests that the common ancestor also intervened *after* conflicts, to assist protagonists in calming down and reconciling. This behavior, seen also in humans, testifies further to the fact that the common ancestor was highly conflict aversive.

A Hypothesis on Moral Origins

An evolutionary definition of morality is behavioral rather than philosophical (Boehm 2000). We must focus on the fact that all human groups reach explicit agreements about behaviors that need to be either encouraged or discouraged and that the values and rules involved are internalized by individuals. In the face of antisocial behaviors these groups will take steps to strongly manipulate the deviants involved and because many types of deviance are aggressive or dangerous, and because measures taken against a deviant may be resisted by his allies, it is necessary for groups to reach unanimity when they decide to sanction him. Indeed, if a minority or bare majority were to sanction a deviant, and his supporters were to step in to defend him, this would throw the group into factionalized conflict far more disruptive than the original deviance (see Boehm 1999a, 2000a).

Elsewhere, in a series of publications (e.g., Boehm 1999b, 2004a,b,c) I have been building a scenario for the evolutionary development of morality sometime after the human line diverged from the *Pan* line. The common ancestral model provides a maximally conservative assessment of key behavioral potentials that were available to subsequent moral evolution: closed hierarchical social groups, formation of larger-than-dyadic political coalitions, and aversiveness to conflict within the group, with active interventions by individuals bent on pacification. As of about seven million years ago, these were the ancient raw materials out of which human moral communities could grow.

Obviously, there is more involved with morality than the formation of large coalitions of group members who share negative feelings toward individuals whose provocative behaviors they wish to discourage. Whether groups could even agree about which behaviors were deviant or laudable without symbolic communication is open to speculation, but de Waal (1996) does provide one anecdote in which this seems to be taking place as captive chimpanzees use a single vocal signal, combined with body language, to express their common hostility toward an individual who is misbehaving.

There are two ways in which the linguistic aspect of culture does seem to be critical to the formation of hunter-gatherer moral communities as we know them. People need to communicate symbolically if they are to agree on values and behavioral standards that make up a well-specified moral code. Perhaps more important, in a fission-fusion group such as a band, in order to identify deviants and agree to collectively manipulate their behavior, highly specific referential communication is needed. This is because many of the individually predatory behaviors that are regularly condemned by foragers can take place in isolation, so victims must be able to report to the group and the group must be able to evaluate the facts. In this context, gossiping seems to be not only an inveterate social fixture that Dunbar (1996) has likened to social grooming, but a generalized means of detecting and evaluating deviance that is crucial to social control as we know it today (Boehm 1999b).

A Hypothesis on the First Sanctioning

Without a common ancestral model, the question of what transgressions earlier humans first defined to be "sins" would be a matter of total speculation. However, if we juxtapose the behaviors that extant hunter-gatherers invariably condemn and sanction with the behavioral repertoire of the common ancestor, there is one (and only one) ancestral behavior that is suggestive as a precursor. Ancestrally, group action was not taken against deception, a rare great ape behavior in any event; nor was it taken against theft, or incest (a favorite candidate of many theorists), or rape. The only ancestral behavior that is directly suggestive of egalitarian moral proscription and sanctioning among hunter-gatherers consists of the efforts of subordinate coalitions to reduce the power of those above them, i.e., to reduce the bullying power of alpha male types.

All four extant hominoids not only show aversion to being dominated (Boehm 1999b), but actively form small coalitions to neutralize such domination. Moralistic humans can carry this much further, to the point that all nomadic foragers remain egalitarian. This means that they treat one another as equals, do not permit bullying behavior that could eventually intimidate an entire group, and even refuse to countenance strong leadership (Boehm 1993). This definitive reversal of the usual dominance process is possible only because entire communities are able to do the necessary political work. Every band is a moral community that insists on egalitarianism and it can unite unanimously to suppress its bullies. This phase of hunter-gatherer moral behavior definitely is presaged in the common ancestor.

There was an ecological development that could have been important in triggering the systematic social control of dangerous deviants by their groups. Humans began to rely increasingly upon large game, and because this was taking place over a long span of evolutionary time, their brains were becoming larger and their ability to gauge the importance of sharing meat (Boehm 1999c, 2000a,c) was becoming quite formidable. There were recurrent junctures at which protracted and heavy reliance upon large mammals was all but dictated by changing meteorological conditions (see Potts 1996), just as is seen with today's Inuit peoples (e.g., Balikci 1970), and at such times it would have become perceptually obvious that a smoothly operating and equitable system for sharing out large-game meat would provide better nutrition and better survival chances for all band members. We call this variance reduction (Smith and Boyd 1990; Kelly 1995), and hunter-gatherers fully appreciate the averaging principles involved (Boehm 2004b).

By contrast, premoral humans surely had a markedly hierarchical social system and therefore were unlikely to engage in efficiently equalized sharing. Indeed, although chimpanzees as dedicated hunters do share meat, it is high-ranking males that get the lion's share (Stanford 1999). Humans are able to share meat quite evenly because social customs and moral rules manipulate the behavior of individuals, and prevent better hunters or bullies from taking disproportionate shares (Kelly 1995).

Early Moral Communities

Once prehistoric bands had learned to work together to suppress alpha tendencies and facilitate the sharing of game, they had arrived at a culturally based methodology that was applicable to other social problems. As cultural animals who also communicated symbolically, referentially, and with displacement (see Gieberman 1998), they were now in a position to create rules in other spheres aside from bullying, and to enforce those rules by means of punishment or social pressure. Included, surely, were rules against cheating, lying, theft, and probably a variety of sex crimes as locally defined. The latter might well have included behaviors that interfered with pair bonding, but there is no way of assuming that the institution of marriage was being supported by earlier moral communities.

In this way, people were able to anticipate and discourage behaviors likely to lead to conflict. When such preemption didn't work, we may assume that like extant foragers, they at least did their best to resolve the conflicts through third-party interventions, be these individual or collective. Although nomadic foragers today keep their leaders too powerless to be effective in mediating really serious conflicts, as groups they are able to manage earlier stages of conflicts fairly effectively, for instance, by staging non lethal duels that dissipate negative energies (see Hoebel 1954).

An Important Residual Problem   The common ancestor was innately prepared to be competitively dominant, and equally good at submitting. It also knew how to join forces politically and to be punitive in working its will, at both individual and group levels. When groups are pushed to the point of moral outrage, punitive social control becomes prominent in all human societies; thus, what was necessary to have moral communities was the ability to set standards of conduct with respect to behaviors deemed to be antisocial and to individually internalize these standards, to identify and communicate about deviants, and to apply punitive collective force in manipulating or eliminating deviants, be the force psychological or physical. One effect of all this was a biologically evolved conscience.

There is an additional aspect of present-day moral communities that is difficult to explain in terms of biologically based precursors. Nomadic hunters go beyond negatively setting up prohibitions and manipulating malefactors by force or threat thereof, for they also issue moralizing calls for altruism and cooperation (Campbell 1972, 1975). These prosocial messages have no precedent in the common ancestor. The tangible rewards include praise, deference, and respect for the virtuous; while this positive sanctioning works in tandem with negative sanctioning, generally it is less reported. This is because it tends to be subtle compared with an enraged community's deciding to criticize, ostracize, expel, or even execute a serious offender.

One explanation for this prosocial side of moral sanctioning, with its frequent calls for willing and generous cooperation, might be that humans have acquired some evolved propensities to altruism that were not present in the common ancestor. At the end of the chapter I will return to positive sanctioning to see how this less appreciated side of moral life fits with the gene–culture coevolutionary scenario I am developing.

The Transition to Morality   I have proposed that important prerequisites for moral behavior were present preadaptively in the common ancestor, at the level of genetic dispositions to dominance, submission, coalition formation, conflict intervention, and tension reduction after conflicts. A subsequent development, not found in the three African apes, was tool use after this capacity led to the invention of hunting weapons suitable for killing conspecifics. This had an equalizing effect on males because projectiles such as spears or clubs could kill at a distance and could be used effectively by men who were less physically powerful (Boehm 1999b). These developments helped to set the stage for political egalitarianism as the first comprehensive moral accomplishment of our species.

It certainly is possible that morality and political egalitarianism developed together gradually, through gene–culture evolution, but it is not easy to imagine what the earlier stages of moral behavior might have been. It seems at least equally possible that a primitive and initially nonmoral kind of egalitarianism was invented rather quickly, with language and the invention of effective hunting weapons as preadaptations that made sudden, planned, and decisive political rebellions relatively safe for the subordinates. If egalitarianism did arrive as an abrupt (and therefore culturally based) change in political format, it is easy to suggest how this new way of doing things socially and politically could have spread through intentional cultural selection (see Boehm 1978, 1982a).

Hunter-gatherers tend to have social intercourse with their neighbors, and with symbolic communication, knowledge of newly egalitarian bands would have been quickly available to members of bands still dominated by alpha types. Thus, even a single successful rebellion of subordination would have made for a demonstration effect, with at least one perceptually obvious benefit: the emergence of personal autonomy for all the family heads in the band. The equalized sharing of large game would have been attractive as well, in case the two were invented simultaneously. I have suggested that this was the case (Boehm 2004a). In either event, the potential for rapid cultural diffusion would have been great, and the frequent long-range migrations forced on most Paleolithic humans by changing climatic cycles (Potts 1996; see also Boehm 1999c) would have greatly assisted the diffusion process.

Could Social Control Have Affected Selection Mechanics?   Based on the best available information and methodology, I have proposed a set of interlocking hypotheses to provide an idea of how a crucial event in human evolution took place and how it was motivated. The rise of morality is of interest, not only because having morals is unique in the animal world, but because moral communities have such special "extra-genetic" ways of inducing behavioral uniformity and are so adept at promoting cooperation. In the remainder of this chapter, I explore both the influence of Late Stone Age human moral communities on the natural selection process and the impact of natural selection on later developments in the field of social control.

The hypothesis is that the advent of egalitarian moral communities of cooperative hunters affected levels of selection in ways that favored the retention of altruistic traits, and that all but definitive moralistic control of serious free riders could have enabled this type of selection to proceed fairly robustly (see Boehm 1997). The first issue is whether moral communities arrived early enough for human gene pools to be significantly changed on this basis.

Surely, all of the ingredients for morally based egalitarianism, including language (Deacon 1997; Lieberman 1998), were available with the appearance of anatomically modern humans. If morality appeared as late as 100,000 years ago, that would provide 4000 generations of natural selection under its special cultural influence. According to E. O. Wilson (1978), a thousand generations would easily be enough for a major trait to evolve, so there is no problem with this chronology. However, to bring needed conservatism to a necessarily speculative task, it is better to place the origin of morality at the point when human cultures began to show elaborate evidence of symbolic thought as other cultural inventions exploded. This brings us to Africa minimally 40,000 years before the present with decorative arts and increasingly complex regional elaborations of stone (and bone) tool technology (see Klein 1999). This would still allow more than the thousand generations that Wilson suggests may be needed. So if moralistic egalitarianism had some influence on natural selection as this changed human nature, there appears to have been time for natural selection to have done its work, even if fully modern style moral communities arrived this late in the anatomically modern human career. There is every reason to believe that the above time estimates are conservative, and that effectively modern moral communities appeared earlier than 50,000 years ago.

Effects of the Egalitarian Syndrome   The prehistoric egalitarian syndrome (see Boehm 1997; Mithen 1990) involved communities in which the heads of households (male or female) morally defined themselves as equals, carefully controlled leadership so it remained weak, and therefore made their decisions by discussing common problems as equals and by trying to find a consensus (see also Boehm 1996). As nomads, the main type of decision they had to make was where their band would migrate when local resources were used up or became scarce. They tried hard to reach a consensus because otherwise the band would have to fragment, and people might find themselves living in groups too small for effective reduction of variance in their meat intake. This means that people cooperated in making decisions because they needed to stay together and cooperate in sharing sporadically acquired large-game meat.

The overall result was a morally based system in which two commodities were always shared within the band. One was political power, which in effect was shared among families. Household heads were equalized in terms of having both personal autonomy in decision making and freedom from being physically or psychologically coerced by others with alpha tendencies. The other commodity was large game, which basically was shared out on an equal basis to every family in the band, just as is done regularly today (see Winterhalder 2001). In all likelihood, possession of mates also was subject to some significant equalization as a side effect of having power and meat distributed equally. The morally backed institution of stable monogamous or polygamous pair bonding (marriage) may have arisen in this context (see Boehm 2000a).

These behaviors, all facilitated by social control, could have had profound effects on the natural selection process (Boehm 1997; see also Boehm 1999a,b,c, 2000a, 2004b). First, by eliminating selfish bullying behavior by alphas, differences over mating opportunity and food quantity and quality were significantly reduced within bands. The result was a drastic reduction in individual phenotypic variation within bands, and therefore a weakening of the selection forces that work against altruism. This reduction took place within bands that, as with extant foraging nomads (see Kelly 1995), surely were composed of a mix of related and unrelated families.

In addition, as an effect of making migration decisions collectively, variance in basic subsistence strategies among families of a band was drastically reduced compared with each family's being on its own. At the within-band level, this further reduced phenotypic variation among individuals or families. In tandem, these two behavior patterns significantly diminished within-group phenotypic variation, even though selection taking place within groups remained a powerful force because extinction rates for individuals were not changed.

In addition, at the between-group level of selection, variation was being enhanced. Because bands usually decided to migrate as units, their subsistence strategies were likely to differ at the band level because each decision was based on a particular concatenation of individuals whose information and manner of processing it were prone to differ. This was particularly the case when ecological stress obliged people to innovate or try risky alternatives, and it was precisely in times of stress that natural selection was likely to operate more powerfully on both individuals and groups. For instance, one band might survive a local drought fairly well by investing their last energy in migrating to an area of possible rainfall, whereas a neighboring band might be seriously decimated because it chose to stay in place and wait it out. The next time around, in an unpredictable environment the advantage might go the other way. It was in this manner that varying strategies among bands amplified phenotypic variation at the between-group level of selection (Boehm 1997). Porous boundaries between bands (see Palmer et al. 1998) did not pose a fatal obstacle to this type of selection process (D. S. Wilson and Sober 1994; see also Sober and Wilson 1998), and with band composition ranging from just a few dozen to perhaps a hundred persons, the limited size of the average units (see Dunbar 1996) was appropriate for some group selection to take place (Wade 1978, see also Bowles et al. 2003). Thus, cooperation among the related and unrelated families living in prehistoric bands could have led to some modest yet behaviorally significant gene selection at the between-group level. The genes involved could have been altruistic (that is, useful to group fitness but deleterious to individual carriers), or selectively neutral for individuals but useful to groups, or useful to both group and individual fitness.

Today, human bands are very good at cooperation. One must question whether this collaboration perhaps has some direct foundation in terms of altruistic genes (see E. O. Wilson 1975), or whether somehow it is accomplished by sociobiologically defined individuals whose motives are essentially selfish and at best nepotistic (e.g., Alexander 1987; Trivers 1971; E. O. Wilson 1978). This second viewpoint still enjoys wide currency. However, even though a great deal of ingenuity has gone into modeling ways in which selfishness (modified only by nepotism) could result in cooperation among nonkin, for humans this position essentially seems to defy common sense.

In reading the hunter-gatherer literature, my impression is that cooperation is in fact likely to be individually ambivalent and that therefore it seems to require some serious backup from social control. However, one must not ignore the positive side of the ambivalence; this could be based on moderate tendencies to altruism. I emphasize this, not only because cooperation often proceeds rather smoothly, but because so often people appear to enjoy engaging in this process and because more basically, individuals seem to readily internalize the prosocial messages that are given in hunting bands. We will return to this issue later.

Some of our greatest scientific achievements do go against common sense, as with the general theory of relativity in physics. In this context, it is tempting to promote a tough-minded kind of sociobiological approach (E. O. Wilson 1975) that defies common sense by reducing all helpfulness and cooperation either to selfishness, to assistance to closer kin, or to carefully metered reciprocation. However, I think it is scientifically useful to give group selection a chance as a controversial theory whose further development has been inhibited by sociobiological and biological biases that in fact require further scrutiny.

Along with extinction rates, it is phenotypic variation that drives natural selection at both individual and group levels. In terms of selection mechanics, I have shown that an egalitarian syndrome (Boehm 1997) influences phenotypic variation in ways that would remain quite invisible to theorists who adhered to the sociobiological habit of doing their modeling at the level of genotype, rather than at the level of phenotype. This widespread practice works well with nonmoral species, but it is risky with one that is capable of decisive social control (e.g., Boyd and Richerson 1992). With a drastic reduction in within-group variation, and with at least a modest increase in between-group variation, these changes at the level of phenotype provide some important food for thought with respect to group selection possibilities.

With moderate group selection in force, minimally, genes that were selectively neutral at the individual level but useful to groups (see D. S. Wilson 1980) would have had a chance of being retained in human gene pools. Here I have in mind aspects of Late Pleistocene cooperation that involved individual contributions that were either cost-free in the first place, or were being evenly reciprocated over time (see Trivers 1971). However, for three decades it has been individually costly altruistic genes that have preoccupied evolutionary biologists, and for their selection there is not only the problem of the lopsided division of labor between within-group and between-group selection, but the formidable obstacle of free riders (see Hamilton 1964; Williams 1966).

With respect to free riders, elsewhere I have made an assessment of the ways in which hunter-gatherers deal with attempts to cheat on systems of cooperation and sharing (Boehm 1999b,c, 2000a, 2004c). I will summarize my findings, but first let me emphasize that on the ground, normal types of sharing among unrelated families within a band are far from being "automatic" and free of contention.

Indeed, the process of sharing out meat often is complicated by elements of selfishness and resentment, and in compensation there are a variety of morally backed cultural institutions that facilitate sharing. Best known are customs that transfer ownership of large carcasses from the proud and predictably arrogant hunters who acquired them to others who are expected to distribute the meat even-handedly (e.g., Lee 1979). Even with this morally based cultural antidote in place, tensions associated with meat sharing are in fact widespread (e.g., Blurton-Jones 1984; Peterson 1993), and the ethnographic record (see Kelly 1995) contains information about occasional sharp quarrels, significant attempts to cheat, and, if rarely, serious punishment of cheaters by the rest of the band. This is suggestive not only of free-riding attempts, but also of control of free riders (Boehm 1999c, 2000a).

Elsewhere (Boehm 1999c) I have predicted that vigilance about the sharing process (see Erdal and Whiten 1994) will become militant, rather than relatively relaxed, under two sets of circumstances. One is when meat becomes scarce, as under conditions of drought or when the migration patterns of prey become less predictable. There are very little ethnographic data on extant foragers who are experiencing serious stress, but the other social accomplishments of hunter-gatherers (see Boehm 1999a) suggest that they are easily sophisticated enough to police free-riding behavior much more carefully during times of scarcity.

The second set of circumstances involves normal times, when basically the common meat supply should be adequate for group needs. If numbers of free-loaders were to rise to such high levels that these social parasites were making just a few hunters do all the work, and were significantly reducing meat intake for the group, this too would be an easily recognized problem. In looking at hunter-gatherer ethnographies, it is clear that under normal circumstances bands occasionally do have an individual who is tempted to take a free ride and manages to do so, for instance, by pretending to be lame when there is hunting to be done. With meat being far from scarce, and with mere suspicions in this direction, the usual reaction is resentment but no active intervention by the group. However, I know of no accounts of bands in which a sizable number of individuals are even suspected of this type of strategy. My prediction is that social tolerance of suspected free riders decreases either when scarcities arise or when the number of "suspects" rises to a level that could affect the diet of others. In a band of thirty or forty persons, this number will not be high.

So hunter-gatherers, while not obsessive in their suppression of free riders, discourage this type of behavior strategically and effectively through social control. In doing so, they make quite modest individual investments in collective acts of social control that can be reproductively costly to the culprit (see D. S. Wilson and Kniffen 1999), and they do this with the most force precisely when their predicted losses will be the most significant. This means that even though free riding is not totally controlled, it is controlled decisively at times when it counts reproductively. This is far from the serious exploitation by free riders that was modeled by Williams (1966) for species lacking social control.

By combining changes in the levels of selection with the effective elimination of free-rider effects, we have a formula for significant group selection of traits useful to groups even if they are moderately altruistic. This process falls within the sociobiological rules set forth by E. O. Wilson (1975), but I must leave it to those who build mathematical models of genetic selection process (e.g., Bowles et al. 2003) to weigh these factors in a more precise manner. For starters, however, I feel strongly that it must be the phenotype and not the genotype that is modeled. It also will be necessary to look closely at the demographic dynamics of bands as group vehicles of selection that have porous boundaries (see D. S. Wilson and Sober 1994), and to consider their size (Wade 1978). It would be necessary, also, to search the ethnographic literature carefully and discern any patterns of splitting and re-formation of hunting bands of the type that Wade (1978) has discussed from the standpoint of laboratory experiments on insects.

Furthermore, even though the contemporary hunter-gatherer literature will be crucial to assessments of prehistoric band composition and size, it will be necessary to take into account not only evolutionary changes in social structure (for example, the nuclear family could be relatively recent), but the vast differences in climatic stability that arrived with the Holocene.

A good example is sex role with respect to hunting. Contemporary ethnographers tend to assume that basically women are not involved in large game hunting, the Agta (Estioko-Griffin 1986) being taken generally as a unique Philippine counterexample.

However, the ethnographic record reveals Bushman wives helping their husbands to track and carry home large game; net hunts by Pygmy females on their own; Peruvian and Paraguayan wives entering into the hunting process; and North American Cree females even hunting moose and bear (Noss and Hewlett 2001). The *Pan*-human ancestor also had females hunting, so the potential for a female hunting role was strongly present in Pleistocene hunting societies.

Why is this fact important? Because it presents new possibilities for modeling group selection in the Pleistocene. The reduction of variance models referred to above posit a minimal hunting band size of about twenty-five, the key factors being that all adult males hunt and that meat be efficiently shared. Five hunters are needed to keep the large game coming in often enough to enable all to receive regular and adequate nourishment (Winterhalder 2001). Bowles et al. (2003) found that twenty-five is about the band size at which groups would have been small enough to fulfill the needs of Wade's (1978) group selection model, while a band size of fifteen would have generated significantly more powerful group selection forces. Imagine, now, a band in which half the women hunt while the other half care for all the children, and you see the minimum viable band size reduced from twenty-five to just over fifteen.

Having half of the band's adult females hunting large game could have been routine, or an emergency measure. If it was routine, this would provide the numbers needed for relatively robust group selection effects. But even if it was an emergency measure, the Pleistocene, with its changeable climates that often put human populations at dire risk, would have provided frequent emergencies. Thus, it can be argued that, whenever bands fell below the threshold of male hunters needed to sustain a regular diet based on large game, females could have filled the gap and humans would have had a shot at surviving in groups as small as fifteen or sixteen.

In working with theoretical or specific case studies, it will be necessary to approximate the changes in the relative strengths of between-group and within-group selection, and, with respect to a given altruistic trait, to calculate the amount of individual cost and the amount of group benefit (see Bowles 2000) and enter these into the equation. However, as a cultural anthropologist I have at least roughed out a research program for the evolutionarily open-minded. What we have here are conditions that favor at least some moderate selection at the between-group level—selection that could support group-beneficial altruistic traits. It was morality, of course, and more specifically egalitarianism and free rider suppression, that made it possible to manipulate a behavioral phenotype so definitively.

Aside from favoring altruistic traits, the arrival of egalitarian moral communities surely had other effects on the natural selection of behavioral dispositions. On the basis of inclusive fitness, it seems likely that humans evolved to be more sensitive to group opinion (see Waddington 1960; Campbell 1975; Boyd and Richerson 1985; Simon 1990). In my opinion, they also evolved to use their individual dominance more cooperatively by acting aggressively as groups to reduce the reproductive advantages of bullies, cheats, and other predatory deviants. After surely several thousand generations of living in moral communities, it is clear that these changes are far from making today's foragers entirely deviance-free or from turning them into a nonhierarchical species, that is, an innately egalitarian species in Vehrencamp's (1983) sense. Hunter-gatherers today still encounter problems with self-aggrandizing selfish upstarts and meet with them quite predictably in spite of an ethos that strongly supports egalitarianism and altruism. What takes place in socially stratified chiefdoms and nations, where social hierarchy is morally countenanced, further demonstrates a hierarchical nature at work (Knauft 1991).

What About Direct Competition Between Groups? The development of a better individual capacity for collaborating aggressively against deviants within the group brought significant benefits of relative fitness to the cooperating moralists, who now suffered far fewer reproductive losses to internal aggressors or cheats. Since this individual capacity for aggressive collaboration was increased by natural selection, the enhancement had one probable side effect: This same capacity can be useful to groups when differences arise between them and they decide to fight.

Did the scale and intensity of warfare rise in earlier or recent times? Despite grossly inadequate Paleolithic evidence (see Keeley 1996), there is at least a suggestive trend in this direction that starts in the Mesolithic; this may have been in part because moral communities preadapted us for warfare. The specific hypothesis would be that as individuals became better adapted to operate as participants in large, assertive moral coalitions in their local hunting communities, a generalized capacity to cooperate in sizable aggressive groups would have been enhanced. This would have lowered the threshold of environmental stimulation needed to produce warfare.

If moral communities and the egalitarian syndrome made possible even some modest selection of altruistic genes, this too could have contributed to the human potential for warfare. This would be true only of intensive warfare, of the type that involves two sizeable groups fighting it out. In contrast to small-scale raiding—in which useful net reproductive gains are derived from cautious, individualistic forays in search of booty—large-scale, intensive armed conflict involves individual patriotic self-sacrifice (see Boehm 1999a; Campbell 1975).

Intensive warfare patterns are in fact endemic among many tribal-agriculturalist peoples (e.g., Meggitt 1977; see also Soltis et al. 1995), and some hunter-gatherers engage in them as well (see Ember 1978). The wonders of modern warfare at the state level are well known, of course, and our capacity to engage in such cooperative high-risk behavior, while it stems in large part from social control, may well be enhanced by prehistoric developments involving natural selection. These involved not only individual participation in cooperatively aggressive efforts to control deviants, mentioned earlier, but also cooperative hunting in small groups that preadapted humans to work together within small military units (see Richerson and Boyd 1999). The natural selection of group-beneficial altruistic traits (Boehm 1999a) could have been important as well. This theory was first advanced by Darwin (1871) himself.

Further Evidence of Altruism   To summarize, once morality arrived in conjunction with political equality, the egalitarian syndrome could have had profound effects on natural selection processes because play was given to group selection forces at the same time that free riding was drastically suppressed. Even though individual extinction rates were not affected, changes in phenotypic variation provided conditions conducive to between-group selection.

At the level of selection mechanics, I believe these arguments to be both logical and plausible. Another type of less direct evidence also supports this same hypothesis. We return to the division of labor between negative and positive types of moral manipulation, as these are seen in hunter-gatherer and all other moral communities. The hypothesis is that the first moral communities relied exclusively on negative sanctioning with groups ganging up to punish miscreants, and that positive sanctioning arrived only after human nature acquired noteworthy altruistic components.

We have seen that members of the earliest moral communities agreed on behaviors they did not like, communicated among themselves as they kept track of deviants, and combined forces to discourage deviant behavior or eliminate serious troublemakers from their groups. Precursors in the common ancestor included a capacity for dominant aggression and a capacity for coalition formation, which in combination made it possible for a group to manipulate (or kill) any of its members. This punitive side of social control has been well recognized since Durkheim (1933) first emphasized it, and more generally the term *social sanctioning* seems to carry a negative connotation of manipulation by threat or punishment (see Boehm 2000a).

This negative type of social reinforcement seems quite consistent with a sociobiological definition of human nature as the product of selection by inclusive fitness: selfish or nepotistic individuals, devoid of altruism, perceive a common threat to their personal welfare in the form of individual deviance and unite to deal with it. At the level of individuals, such cooperation is reproductively rewarded because each person shares a major fitness dividend that comes from getting rid of bullies or other social predators within the group, and because individual risks and energy spent in collective sanctioning are modest. The deviant loses and the rule-enforcing citizens share the net profit. Negative sanctioning of this type would seem to be adequate to the task of cutting down political upstarts and maintaining an egalitarian order, and it can be hypothesized that early moral communities operated in this way. The same aggressive techniques could have been applied in controlling other types of predators within the band and in policing the equalized sharing of large game.

If we move up to the present, we see a continuation of this punitive approach but also what might be called positive sanctioning. There are two questions that I wish to explore here. First, is a wholly selfish species likely to come up with calls for generosity and cooperation when it already has at its disposal some effective means of negative reinforcement? Second, are members of a wholly selfish species likely to be responsive to such positively phrased messages?

By "wholly selfish" I refer to a species whose genotype is shaped solely by within-group selection or inclusive fitness. It is conceivable that a wholly selfish species, one that understands the individual dividends that come from cooperation, might consider adding the use of carrots to their successful use of sticks to make cooperation work still better. This cannot be ruled out. However, even a modest element of altruism in human nature would make positive reinforcement far more likely to work, for in a multifamily band, even an ambivalently altruistic individual is more likely to respond to a social message that calls for generous contributions to group efforts. By the same token, an ambivalently altruistic being is more likely to come up with such prosocial messages in the first place than is a wholly selfish being.

One must keep in mind not only that the evolved altruistic component in human nature is likely to be relatively modest but also that at the level of the phenotype, feelings of generosity toward nonkin are heavily amplified by the socialization of children in prosocial directions (see Goody 1991) and by social rewards that stimulate generous cooperation in adults. This results in the internalization of prosocial values.

The result is not perfectly motivated cooperation, but effective and usually willing collaboration that invariably reduces variance in meat supply when large game is being acquired sporadically. On the one hand, today's hunter-gatherers do show evidence of ambivalence about sharing their large game. On the other hand, basically the job of reducing variance is accomplished quite efficiently (see Kelly 1995) in spite of occasional complaints and squabbles.

In this context, the internalization of values that favor generosity and cooperation can be quite strong. Indeed, the tendency to share meat when others (outside the family) request it often seems to be all but automatic, and Kelly (1995) provides an anecdote in which the hunter-gatherer sharing ethic is so well internalized that to deny such a request is virtually unthinkable.

It is of interest that socially instigated calls for generosity and cooperation—the ones identified by Campbell (1965a, 1972) and found in all six early human civilizations (Campbell 1975)—seem to be echoed in every type of human society. There is no way to prove that individuals who were entirely "selfish" sociobiologically couldn't have brilliantly come up with such proclamations as manipulative devices, acting purely from Machiavellian motives. However, these calls for generosity and willing cooperation are more logically explained if we cede to human nature a modicum of innate altruism that can be heavily amplified by cultural conditioning.

I have suggested that the first human moral communities were likely to have dealt in negative reinforcement only; power plays by large, morally aggressive coalitions were the name of the game. However, as the egalitarian syndrome prevailed over hundreds and hundreds of generations, and altruistic traits were subjected to at least some moderate positive selection at the between-group level, human nature was being modified. Once altruistic traits were in place and were strong enough to make a difference behaviorally, moral life acquired this new, positive component that we see in all human moral systems today.

Summary

In considering the interactions of culture and natural selection in the Upper Paleolithic, I have hypothesized that there were a number of genotypically well-prepared precursor behaviors that preadapted our common ancestor for the evolutionary emergence of egalitarian social communities as a unique human invention. As a cultural invention, the acquisition of lethal hunting weapons also made a contribution to this development. The result was the egalitarian syndrome, which was possible only with communities that were judgmental and capable of collective sanctioning that was guided by human intentions (Boehm 1999b, 2000a).

At first social control was likely to have been based on threat or the use of force, and therefore it was merely punitive. Nothing more was needed to produce the egalitarian syndrome that rearranged the division of labor between between-group and within-group selection and saw the effective curbing of free riders through group sanctioning. The conscience and positive social sanctioning came later (Boehm 2000a).

When punitive sanctioning was complemented later on by positive moral manipulations, this new development was likely to have depended upon the addition of altruistic tendencies to human nature, tendencies that became strongly reinforced by socialization and group opinion. The final result was a group in which positive and negative types of social reinforcement worked in tandem, and both types of manipulation supported cooperation as a consciously directed human activity.

With respect to issues of coevolution (e.g. Durham 1991), the scenario I have developed involves natural selection and cultural processes in several ways. First, selection at the biological level set in place the precursors to moral behavior that were conservatively identified in the common ancestor. Next, it would appear that egalitarian bands appeared as a product of intentional cultural invention, with subordinates realizing that by standing together they could become individually autonomous as they shared large game. In being cultural this process could have been protracted or quite rapid. Next, the resulting egalitarian syndrome profoundly affected natural selection processes, giving significant play to group selection. Finally, evolved altruistic tendencies began to influence social control practices providing a positive cast to sanctioning that complemented—but did not replace—the negative orientation of earlier moral communities.

These hypotheses are never likely to be developed to a degree that outright falsification is likely. However, their importance for defining the human condition justifies their being considered on the basis of relative plausibility. As a cultural anthropologist I have not attempted to create mathematical models to test the important hypothesis about group selection, but I hope the scenario I have developed may inspire others to do further work in this direction. I expect that it may be possible to create realistic models of prehistoric multifamily hunting bands, taking into account their size, their considerable instability, the ways in which they grew or shrank as Pleistocene Climates varied, the ways in which they fissioned, the ways in which new bands were formed, and the possible role of females in hunting. This would be helpful in further evaluating the relative plausibility of the group-selection arguments I have developed.

Acknowledgments

I wish to thank the Fyssen Foundation for inviting me to the Paris conference on which this book is based and the Templeton Foundation for making possible 3 years of research on hunter-gatherer conflict resolution. I also wish to thank the H. F. Guggenheim Foundation for two grants that supported research on, respectively, egalitarian behavior of humans and conflict intervention in chimpanzees. In addition, a fellowship at the School of American Research in Santa Fe, New Mexico and a Simon Guggenheim research fellowship made it possible to develop some of the newer arguments in this chapter. I also thank the Santa Fe Institute for providing a forum for developing the hypothesis in this chapter.

References

Aureli, F., and de Waal, F. B. M. (eds.) 2000. *Natural conflict resolution.* Berkeley: University of California Press.

Alexander, R. D. 1987. *The biology of moral systems.* New York: Aldine de Gruyter.

Balikci, A. 1970. *The Netsilik Eskimo.* Prospect Heights, Ill.: Waveland.

Blurton-Jones, N. G. 1984. A selfish origin for human food sharing: Tolerated theft. *Ethology and Sociobiology* 4: 145–147.

Boehm, C. 1978. Rational preselection from *Hamadryas* to *Homo sapiens*: The place of decisions in adaptive process. *American Anthropologist* 80: 265–296.

Boehm, C. 1982a. A fresh outlook on cultural selection. *American Anthropologist* 84: 105–124.

Boehm, C. 1982b. The evolutionary development of morality as an effect of dominance behavior and conflict interference. *Journal of Social and Biological Sciences* 5: 413–422.

Boehm, C. 1984. Can hierarchy and egalitarianism both be ascribed to the same causal forces? *Politics and the Life Sciences* 1: 34–37.

Boehm, C. 1991. Lower-level teleology in biological evolution: Decision behavior and reproductive success in two species. *Cultural Dynamics* 4: 115–134.

Boehm, C. 1993. Egalitarian society and reverse dominance hierarchy. *Current Anthropology* 34: 227–254.

Boehm, C. 1994. Pacifying interventions at Arnhem Zoo and Gombe. In R. W. Wrangham, W. C. McGrew, F. B. M. de Waal, and P. G. Heltne (eds.), *Chimpanzee cultures* (pp. 211–226). Cambridge, Mass.: Harvard University Press.

Boehm, C. 1996. Emergency decisions, cultural selection mechanics, and group selection. *Current Anthropology* 37: 763–793.

Boehm, C. 1997. Impact of the human egalitarian syndrome on Darwinian selection mechanics. *American Naturalist* 150: 100–121.

Boehm, C. 1999a. Forager hierarchies, innate dispositions, and the behavioral reconstruction of prehistory. In M. W. Diehl (ed.), *Hierarchies in action: Cui bono?* Center for Archaeological Investigations, Occasional Paper No. 27. (pp. 31–58). Carbondale: Southern Illinois University Press.

Boehm, C. 1999b. *Hierarchy in the forest*. Cambridge, Mass.: Harvard University Press.

Boehm, C. 1999c. The natural selection of egalitarian traits. In C. Boehm (guest editor), *Human Nature*. Special issue on group selection. 10: 205–252.

Boehm, C. 2000a. Conflict and the evolution of social control. In L. Katz (guest editor), *Journal of Consciousness Studies*, special issue on evolutionary origins of morality 7: 79–183.

Boehm, C. 2000b. Group selection in the Upper Paleolithic. In L. Katz (guest editor), *Journal of Consciousness Studies*, Special issue on the evolutionary origins of morality. 7: 1–2: 211–215.

Boehm, C. 2004a. Large game hunting and the evolution of human sociality. In R. W. Sussman and A. Chapman (eds.), *The origins and nature of sociality* (pp. 270–287). New York: Aldine de Gruyter.

Boehm, C. 2004b. Variance-reduction and the evolution of social control. Paper presented at symposium on co-evolution of behaviors and institutions, Santa Fe Institute. (Santa Fe Institute website pdf)

Boehm, C. 2004c. What makes humans economically distinctive? A three-species evolutionary comparison and historical analysis. *Journal of Bioeconomics* 2: 109–135.

Bowles, S. 2000. Individual interactions, group conflicts, and the evolution of preferences. Working papers, Santa Fe Institute. In S. Durlauf and H. P. Young (eds.), *Social Dynamics* (pp. 155–190). Cambridge, Mass.: MIT Press.

Bowles, S., Choi, J.-K., and Hopfensitz, A. 2003. The co-evolution of individual behaviors and social institutions. *Journal of Theoretical Biology* 223: 135–147.

Boyd, R., and Richerson, P. J. 1982. Cultural transmission and the evolution of cooperative behavior. *Human Ecology* 10: 325–351.

Boyd, R., and Richerson, P. J. 1985. *Culture and the evolutionary process.* Chicago: University of Chicago Press.

Boyd, R., and Richerson, P. J. 1991. Culture and cooperation. In R. A. Hinde and J. Grobel (eds.), *Cooperation and prosocial behavior* (pp. 27–48). Cambridge: Cambridge University Press.

Boyd, R., and Richerson, P. J. 1992. Punishment allows the evolution of cooperation (or anything else) in sizable groups. *Ethology and Sociobiology* 13: 171–195.

Campbell, D. T. 1965a. Ethnocentric and other altruistic motives. In D. Levine (ed.), *Nebraska symposium on motivation* (pp. 283–311). Lincoln: University of Nebraska Press.

Campbell, D. T. 1965b. Variation and selective retention in socio-cultural evolution. In H. R. Barringer, B. I. Blanksten, and R. W. Mack (eds.), *Social change in developing areas* (pp. 19–49). Cambridge: Schenkman.

Campbell, D. T. 1972. On the genetics of altruism and the counter-hedonic component of human culture. *Journal of Social Issues* 28: 21–37.

Campbell, D. T. 1975. On the conflicts between biological and social evolution and between psychology and moral tradition. *American Psychologist* 30: 1103–1126.

Cavalli-Sforza, L. L., and Feldman, F. W. 1981. *Cultural transmission and evolution: A quantitative approach.* Princeton, N.J.: Princeton University Press.

Cashdan, E. A. 1980. Egalitarianism among hunters and gatherers. *American Anthropologist* 82: 116–120.

Cashdan, E. A. 1983. Territoriality among human foragers: Ecological models and an application to four Bushman groups. *Current Anthropology* 24: 47–66.

Cords, M. 1997. Friendships, alliances, reciprocity and repair. In D. Byrne and A. Whiten (eds.), *Machiavellian intelligence II* (pp. 24–49). Cambridge: Cambridge University Press.

Darwin, Charles. 1871. *The descent of man and selection in relation to sex.* London: John Murray.

Deacon, T. W. 1997. *The symbolic species: The co-evolution of language and the brain.* New York: Norton.

Dunbar, R. 1996. *Grooming, gossip and the evolution of language.* London: Faber and Faber.

Durham, W. H. 1976. The adaptive significance of cultural behavior. *Human Ecology* 4: 89–121.

Durham, W. H. 1982. Interactions of genetic and cultural evolution: Models and examples. *Human Ecology* 10: 289–323.

Durham, W. H. 1991. *Coevolution: Genes, culture, and human diversity*. Stanford, Calif.: Stanford University Press.

Durkheim, E. 1933. *The division of labor in society*. New York: Free Press.

Dyson-Hudson, R., and Smith, E. A. 1978. Human territoriality: An ecological reassessment. *American Anthropologist* 80: 21–41.

Ember, C. 1978. Myths about hunter-gatherers. *Ethnology* 17: 439–448.

Erdal, D., and Whiten, A. 1994. On human egalitarianism: An evolutionary product of Machiavellian status escalation? *Current Anthropology* 35: 175–184.

Erdal, D., and Whiten, A. 1996. Egalitarianism and Machiavellian intelligence in human evolution. In P. Mellars and K. Gibson (eds.), *Modelling the early human mind* (pp. 139–150). Cambridge: MacDonald Institute for Archeological Research.

Estioko-Griffin, A. 1986. Daughters of the forest. *Natural History* 95: 36–43.

Flanagan, J. G. 1989. Hierarchy in simple "egalitarian" societies. *Annual Review of Anthropology* 18: 245–266.

Foley, R. 1997. *Humans before humanity*. Oxford: Blackwell.

Fossey, D. 1983. *Gorillas in the mist*. Boston: Houghton-Mifflin.

Fried, M. H. 1967. *The Evolution of political society: An essay in political anthropology*. New York: Random House.

Furer-Haimendorf, C. von. 1967. *Morals and merit: A study of values and social controls in South Asian societies*. Chicago: University of Chicago Press.

Gardner, P. 1991. Foragers" pursuit of individual autonomy. *Current Anthropology* 32: 543–558.

Goodall, J. 1979. Life and death at Gombe. *National Geographic* 155: 592–622.

Goodall, J. 1986. *The Chimpanzees of Gombe*. Cambridge, Mass.: Harvard University Press.

Goody, E. 1991. The learning of prosocial behavior in small-scale egalitarian societies: An anthropological view. In R. A. Hinde and J. Grobel (eds.), *Cooperation and prosocial behavior* (pp. 106–128). Cambridge: Cambridge University Press.

Hamilton, W. D. 1964. The genetical evolution of social behavior I, II. *Journal of Theoretical Biology* 7: 1–52.

Hoebel, E. A. 1954. *The law of primitive man: A study in comparative legal dynamics*. Cambridge, Mass.: Harvard University Press.

Kano, T. 1992. *The last ape: Pygmy chimpanzee behavior and ecology*. Stanford, Calif.: Stanford University Press.

Keeley, L. H. 1996. *War before civilization: The myth of the peaceful savage*. New York: Oxford University Press.

Kelly, R. L. 1995. *The foraging spectrum: Diversity in hunter-gatherer lifeways*. Washington, D.C.: Smithsonian Institution Press.

Klein, R. G. 1999. *The human career: Human biological and cultural origins*. Chicago: University of Chicago Press.

Knauft, B. B. 1991. Violence and sociality in human evolution. *Current Anthropology* 32: 391–428.

Knauft, B. B. 1994. Reply to Erdal and Whiten. *Current Anthropology* 35: 181–182.

Lee, R. B. 1979. *The !Kung San: Men, women, and work in a foraging society*. Cambridge: Cambridge University Press.

Lieberman, P. 1998. *Eve spoke: Human language and human evolution*. New York: Norton.

Meggitt, M. 1977. *Blood is their argument*. Palo Alto, Calif.: Mayfield.

Mithen, S. J. 1990. *Thoughtful foragers: A study of prehistoric decision making*. Cambridge: Cambridge University Press.

Nishida, T. 1979. The social structure of chimpanzees of the Mahale Mountains. In D. A. Hamburg and E. R. McCown (eds.), *The great apes* (pp. 73–122). Menlo Park, Calif.: Benjamin/Cummings.

Noss, A. J., and Hewlett, B. S. 2001. The contexts of female hunting in Central Africa. *American Anthropologist* 103: 1024–1040.

Palmer, C. T., Fredrickson, B. E., and Tilly, C. F. 1998. Categories and gatherings: Group selection and the mythology of cultural anthropology. *Ethology and Sociobiology* 18: 291–308.

Peterson, N. 1993. Demand sharing: Reciprocity and the pressure for generosity among foragers. *American Anthropologist* 95: 860–874.

Potts, R. 1996. *Humanity's descent: The consequences of ecological instability*. New York: Avon.

Richerson, P. J., and Boyd, R. 1999. Complex societies: The evolutionary origins of a crude super-organism. In C. Boehm (guest editor), *Human Nature*. Special issue on group selection. 10: 253–289.

Richerson, P. J., and Boyd, R. 2005. *Not by genes alone: How culture transformed human evolution*. Chicago: University of Chicago Press.

Service, E. R. 1962. *Primitive social organization: An evolutionary perspective*. New York: Random House.

Service, E. R. 1975. *Origin of the state and civilization: The process of cultural evolution*. New York: Norton.

Simon, H. 1990. A mechanism for social selection and successful altruism. *Science* 250: 1665–1668.

Smith, E. A. and Boyd, R. 1990. Risk and reciprocity: Hunter-gatherer socioecology and the problem of collective action. In E. A. Cashdan (ed.), *Risk and uncertainty in tribal and peasant economies* (pp. 167–192). Boulder, Col.: Westview Press.

Sober, E., and Wilson, D. S. 1998. *Unto others: The evolution and psychology of unselfish behavior.* Cambridge, Mass.: Harvard University Press.

Soltis, J., Boyd, R., and Richerson, P. J. 1995. Can group-functional behaviors evolve by cultural group selection? An empirical test. *Current Anthropology*, 36: 473–494.

Stanford, C. B. 1998. The social behavior of chimpanzees and bonobos: Empirical evidence and shifting assumptions. *Current Anthropology* 14: 399–420.

Stanford, C. B. 1999. *The hunting apes: Meat eating and the origins of human behavior.* Princeton, N.J.: Princeton University Press.

Tattersall, I. 1993. *The human odyssey: Four million years of human evolution.* New York: Macmillan.

Tiger, L., and Fox, R. 1971. *The imperial animal.* New York: Delta.

Trivers, R. L. 1971. The evolution of reciprocal altruism. *Quarterly Review of Biology* 46: 35–57.

Vehrencamp, S. L. 1983. A model for the evolution of despotic versus egalitarian societies. *Animal Behavior* 31: 667–682.

de Waal, F. B. M. 1982. *Chimpanzee politics: Power and sex among apes.* New York: Harper and Row.

de Waal, F. B. M. 1989. *Peacemaking among primates.* Cambridge, Mass.: Harvard University Press.

de Waal, F. B. M. 1996. *Good natured: The origins of right and wrong in humans and other animals.* Cambridge, Mass.: Harvard University Press.

Waddington, C. H. 1960. *The ethical animal.* Chicago: University of Chicago Press.

Wade, M. J. 1978. A critical review of the models of group selection. *Quarterly Review of Biology* 53: 101–114.

Wiessner, P. 1996. Leveling the hunter: Constraints on the status quest in foraging societies. In P. Wiessner and W. Schiefenhovel (eds.), *Food and the status quest: An interdisciplinary perspective* (pp. 171–191). Oxford: Berghahn.

Wiessner, P. 1998. On emergency decisions, cultural selection mechanics, and group selection. *Current Anthropology* 39: 356–358.

Williams, G. C. 1966. *Adaptation and natural selection: A critique of some current evolutionary thought.* Princeton, N.J.: Princeton University Press.

Wilson, D. S. 1980. *The natural selection of populations and communities.* Menlo Park, Calif.: Cummins.

Wilson, D. S., and Kniffin, K. M. 1999. Multilevel selection and the social transformation of behavior. In C. Boehm (guest editor), *Human Nature*, Special issue on group selection. 10: 291–310.

Wilson, D. S., and Sober, E. 1994. Reintroducing group selection to the human behavioral sciences. *Behavioral and Brain Sciences* 17: 585–654.

Wilson, E. O. 1975. *Sociobiology: The new synthesis.* Cambridge, Mass.: Harvard University Press.

Wilson, E. O. 1978. *On human nature.* Cambridge, Mass.: Harvard University Press.

Winterhalder, B. 2001. Intragroup resource transfers: Comparative evidence, models, and implications for human evolution. In C. B. Stanford and H. T. Bunn (eds.), *Meat-eating and human evolution* (pp. 279–301). Oxford: Oxford University Press.

Woodburn, J. 1979. Minimal politics: The political organization of the Hazda of North Tanzania. In W. A. Shack and P. S. Cohen (eds.), *Politics in leadership: A comparative perspective* (pp. 244–266). Oxford: Clarendon Press.

Woodburn, J. 1982. Egalitarian societies. *Man* 17: 431–451.

Wrangham, R. 1987. African apes: The significance of African apes for reconstructing social evolution. In W. G. Kinzey (ed.), *The Evolution of human behavior: Primate models* (pp. 51–71). Albany: State University of New York Press.

Wrangham, R., and Peterson, D. 1996. *Demonic males: Apes and the origins of human violence.* Boston: Houghton-Mifflin.

5 Solving the Puzzle of Human Cooperation

Robert Boyd and Peter J. Richerson

No Consensus on Cooperation

Is society an organic whole, with each of its many components working together like the organs in a body? Like organisms, societies are composed of many parts that seem to work together to enhance their survival. Different people fulfill different, necessary roles—providing subsistence, reproduction, coordination, and defense. Regular exchange of matter and energy guarantees that each component has the resources it needs. Norms, laws, and customs regulate virtually every aspect of social interaction: who may marry whom, how disputes are resolved, and how verbs should be conjugated. Ritual and religion provide comfort to the sick and fearful, maintain a feeling of solidarity and belonging, and serve to preserve and transmit knowledge through time. Even the simplest human societies seem like complex machines designed for growth and survival.

People have long been divided about whether this metaphor is useful or misleading. Many believe that the appearance of design is real. Functionalism, an old and still influential school in anthropology and sociology, holds that beliefs, behaviors, and institutions exist because they promote the healthy functioning of social groups (Spencer 1891; Radcliffe-Brown 1952; Malinowski 1922; Aberle et al. 1950). Such functionalists believe that the existence of some observed behavior or institution is explained if it can be shown how the behavior or institution contributes to the health or welfare of the social group. The conviction that people are selfish drives others to argue that the appearance of design is an illusion; the complex structure merely reflects a standoff in a struggle among selfish individuals. Such rational individualists, mainly economists, political scientists, and philosophers, hold that human choices must be explained in terms of individual benefits; any group benefits are an accidental side effect of selfish individual choices.

This conflict remains unresolved because the competing protagonists espouse irreconcilable views about the causes of human action. Functionalists view people as being shaped by their society. People acquire a belief in the rightness of the norms and customs of their culture as a result of growing up in that society. Believers in rational individualism see people as choosing how to behave based on their own interest. People are bound by custom only to the extent that the custom serves these interests. Let us now consider in more detail why this difference in belief about human nature leads to different views about the function of society.

When functionalists do provide a mechanism for the generation or maintenance of group-level adaptations, it is usually in terms of selection among social groups (Turner and Maryanski 1979). Rappaport provides an exceptionally clear statement of the idea:

[S]uch conceptions as honor, morality, altruism, honesty, valor, righteousness, prestige, gods, heaven, and hell, [make] group selection important among humans. By group selection I mean the selection for and perpetuation of conventions enhancing the persistence of groups, even though these conventions can be disadvantageous to those individuals whose actions accord most closely with them. (Rappaport 1984: 401)

Suppose that societies have a member of functional prerequisites without which they have difficulty surviving. Social groups whose culturally transmitted behaviors, beliefs, and institutions do not provide these prerequisites frequently become extinct, leaving only those societies with functional cultural attributes as survivors. On this argument, selection among social groups causes societies to be adapted to their circumstances in much the same way that selection among individuals causes organisms to be adapted to their environments. We refer to this process as cultural group selection because it involves the differential survival and proliferation of culturally variable groups.

Most rational individualists have paid little attention to the group selection argument advanced by functionalists. The reason, we think, is captured by Margaret Thatcher's well-known (famous on the right , infamous on the left) aphorism, "There is no such thing as society." There cannot be selection among societies because societies do not have intrinsic properties; people make choices in their own interest. If two societies are different, then there must be some difference in conditions that causes individuals to choose differently. Societies cannot replicate themselves because their properties depend on external conditions, not on transmissible properties. A small minority of rational individualists have taken an interest in selection among groups, and as we will see later in this chapter, their ideas provide one way of bridging the gap between functional- and individual-level explanations.

Rational individualists have devoted considerable ingenuity to devising selfish explanations for the seemingly group-beneficial features of society. The first and most famous of these is Adam Smith's invisible hand. Smith showed how the market could regulate the flow of goods and services in a complex society so that all the things necessary to sustain a city like London are supplied each day without planning. In recent years, scholars have extended this kind of reasoning to many kinds of social interactions that are not mediated by markets (Becker 1981; Coleman 1990; Young 1998).

Many of these authors believe that all of the complexity of human cooperation can be explained solely in terms of individual self-interest.

Sociobiology Deepens the Puzzle

Placing the question in an evolutionary framework deepens the puzzle because it supports the rational individualists' assumption that people are selfish, but it casts serious doubt on the rational individualists' conclusion that selfish individuals will form complex cooperative societies.

The proposition that human behavior is a product of organic evolution strongly supports the view that people are selfish. Evolutionary theory predicts that any heritable tendency to behave altruistically toward nonrelatives will be rapidly eliminated by natural selection. To see why, suppose that some individuals in a population have a heritable tendency to help other, unrelated members of their social group at a cost to themselves. For example, suppose some females were motivated by generalized maternal feelings to suckle the orphaned offspring of other females. Such compassionate females would have fewer offspring on average than females who lacked this propensity because the compassionate females would have less milk for their own offspring, and all other things being equal, this would reduce their offspring's survival. Thus, in each generation there will be fewer copies of the genes that create the motive to suckle orphans and eventually the tendency will disappear.

Selection will favor selfless behavior in only one circumstance: when it is directed toward genetic relatives. To see why, suppose that some females have a heritable tendency to suckle a sister's offspring when they are in need. Since such offspring have a 50 percent chance of carrying the same genes as the female's own offspring, selection will usually favor such nepotistic motives if the increase in fitness of the sister's offspring is more than twice the reduction in fitness of the female's own offspring. This reasoning, first elaborated by W. D. Hamilton (1964), is supported by an immense body of field and laboratory observation and measurement. It is certainly possible that humans are unusual in some way that caused them to evolve unselfish motives. However, the burden of proof is on people taking this view to show exactly why humans are odd, and in the absence of a clear demonstration of why we are odd, the straightforward prediction of evolutionary biology is that human actions result from selfish or nepotistic motives.

In other species, complex cooperative societies exist only when their members are close relatives. In most animal species, cooperation is either limited to very small groups or is absent altogether. Among the few animals that cooperate in large groups are social insects such as bees, ants, and termites, and the naked mole rat, a subterranean African rodent. Multicellular plants and many forms of multicellular invertebrates can also be thought of as eusocial societies made up of individual cells. In each of these cases, the cooperating individuals are closely related. The cells in a multicellular organism are typically members of a genetically identical clone, and the individuals in insect and naked mole rat colonies are siblings.

Evolutionary biologists believe that complex cooperative systems are limited to societies of relatives because such systems are vulnerable to self-interested cheating. The many members of an ant colony cannot easily monitor the behavior of all the other members; thus each member has the opportunity to cheat on the system. For example, rather than maintain the colony and feed the queen's offspring, a worker termite can devote time and energy to laying her own fertile eggs. Since the colony has many members, the effect of each on the functioning of the whole group is small and therefore each is better off if he or she cheats. A division of labor creates further opportunities for cheating because it requires exchanges of "goods and services" whose provision is separated in time.

In contrast to the societies of other animals, virtually all human societies are based on the cooperation of large numbers of unrelated people. This is obviously true of modern societies in which complex tasks are managed by enormous bureaucracies such as the military, political parties, churches, and corporations. Markets coordinate the activity of millions of people and allow astonishing specialization. This is also true of the societies that have characterized the human species since the first intensive broad-spectrum foraging, and later agriculture, allowed sedentary settlements. Consider, for example, the societies of highland New Guinea. Here, patrilineally organized groups number a few hundred to several thousand. These groups have religious, political, and economic specialists; they engage in trade and elaborate ritual exchange with distant groups; and they are able to regularly organize parties numbering several hundred to make war on their neighbors. Even contemporary hunter-gathers who are limited to the least-productive parts of the globe have extensive exchange networks and regularly share food and other important goods outside the family. Other animals do none of these things.

Thus we have an evolutionary puzzle. Our Miocene primate ancestors presumably cooperated only in small groups that were mainly made up of relatives, like those of contemporary nonhuman primates. Such social behavior is consistent with our understanding of how natural selection shapes behavior. Over the next five to ten million years, something happened that caused humans to cooperate in large groups. The puzzle is, what caused this radical divergence from the behavior of other social mammals? Did some unusual evolutionary circumstance cause humans to be less selfish than other creatures? Or do humans have some unique feature that allows them to better organize complex cooperation among selfish nepotists?

Solutions to the Puzzle   Researchers have proposed five different kinds of solutions to this puzzle:

1. The "heart on your sleeve" hypothesis holds that humans are cooperative because they can truthfully signal cooperative intentions.
2. The "big mistake" hypothesis proposes that contemporary human cooperation results from psychological predispositions that were adaptive when humans lived in small groups of relatives.
3. Manipulation hypotheses hold that people are either tricked or coerced into cooperating in the interest of others.
4. Moralistic reciprocity hypotheses hold that greater human cognitive abilities and human language allow humans to manage larger networks of reciprocity, which accounts for the extent of human cooperation.
5. Cultural group selection hypotheses argue that the importance of culture in determining human behavior causes selection among groups to be more important for humans than for other animals.

These five are not mutually exclusive and in fact we believe that the most likely explanation is some combination of the last two hypotheses.

The "Heart on Your Sleeve" Hypothesis

In his book, *Passions within Reason*, the economist Robert Frank (1988) argues that humans cooperate with nonrelatives because people can reliably detect the true intentions of others. Frank thinks that people have innate dispositions that cause them to be more or less cooperative and that they can signal their disposition to others by their appearance and demeanor. Truthful signals allow cooperators to preferentially interact with other cooperators, and this makes possible the evolution of cooperative behavior among nonrelatives. Because thinking and feeling are complex physiological processes, Frank argues, it is more costly for a defector to maintain the appearance of a nice guy than it is for a cooperator. The theory of signaling developed in economics and borrowed by biologists (see Gintis 2000 for a discussion) shows that with this assumption, natural selection may favor honest signals of intent. Thus, Frank's idea is that we wear our hearts, be they good or bad, on our sleeves.

This argument cannot explain why humans are more cooperative than other animals because it applies with equal force to other animals. If humans must wear their hearts on their sleeves, then why not chimps or baboons? True, humans seem smarter than other creatures (although this may mainly be a matter of perspective), but this fact cuts both ways. Being clever may allow humans to better penetrate deceptions, but it may also allow them to better perpetrate deceptions as well. In fact, we would use the evidence from other animals to argue against the importance of Frank's mechanism among humans. If the evidence about intentions that one could obtain by the mechanisms that Frank invokes was as reliable as the information conveyed by kinship, then we should see much cooperation among nonrelatives in nature. Since we see very little cooperation among nonrelatives, we conclude that signaling provides much poorer information than kinship. Perhaps the reason is that the costs of a defector deceptively signaling that they are altruists is really not very high. If deceptive signalers of altruism can easily reap most of the benefits provided by true altruists without paying the costs of providing altruism, the signal will become useless.

The "Big Mistake" Hypothesis

Many people have argued that contemporary cooperation results from dispositions that evolved when humans lived in small groups of close relatives (e.g., Alexander 1974, 1987; Hamilton 1975; Tooby and Cosmides 1989). The genus *Homo* is about two million years old, and the species *Homo sapiens* at least 160,000 years old. Until the spread of agriculture, beginning about 11,500 years ago, humans probably lived in relatively small groups (although how small is a matter of debate). In such a world, it is argued, selection could favor psychological mechanisms that led to unselective altruism toward fellow group members because all the potential recipients were relatives. With the advent of agriculture and later urban life, the size of human groups increased dramatically, and these same predispositions led to altruism toward nonrelatives. For example, most people feel sympathy for others. Few can fail to be moved by a photograph of a starving Somali child, its eyes rimmed with flies. Such feelings cause many people to act, to send money, to organize relief, or to choose a less than lucrative career in international medicine or development. Sympathy of this kind is genetically maladaptive in the contemporary world because it leads to indiscriminate altruism, but it would have been fine, the argument runs, when all of the potential recipients were relatives. We call this the big mistake hypothesis because almost everything in modern life—trade, religion, government, and science—is a mistake from the genes' point of view.

The big mistake hypothesis can be extended to explain intergroup enmity (van den Berghe 1981). In many species of nonhuman primates, kinship ties within groups are stronger than kinship ties among groups and as a result neighboring groups compete for territory or other resources. In such an environment, selection might favor a generalized enmity toward nongroup members. Today, neighboring groups of hunters are typically physically indistinguishable but often have different dialects, customs, and styles of manufacturing items such as arrow points. Thus, it is argued, selection could have favored the rule: "Be nice to people who talk like you, dress like you, and act like you. Be nasty to everyone else."

The big mistake hypothesis is a cogent explanation, but we think the evidence from the study of nonhuman primates works against it. Many contemporary nonhuman primates live in small, kin-based groups very much like those posited for early humans. However, life in such groups does not show much nonselective altruism. In all primate species, the members of at least one sex leave their natal group and join another group, where their only relatives will be their own children. In most species, it is the males who leave, but there are some species, such as chimpanzees, in which females emigrate. Members of the other sex, usually females, remain in their natal group and live among their relatives. However, when groups are of any size, some of these relatives will be quite close whereas others will be quite distant.

A great deal of evidence suggests that primates are very sensitive to differences in relatedness, directing costly helping behavior mainly toward relatives. Consider baboons, for example. A typical baboon group might consist of about sixty animals, about half of them infants and juveniles, four or five adult male immigrants, and the rest adult females born in the group. Violent conflict among females occurs hourly, and these conflicts sometimes result in serious injuries. Many conflicts involve coalitions, which are usually made up of matrilineal kin. As a result, female baboons are organized into a dominance hierarchy in which all members of a matriline occupy adjacent positions. Violent conflict among males is also common, but coalitions are much rarer (they are even absent in the chacma baboon) and more ephemeral. When the dominant male confronts a subordinate, the subordinate will often grab an infant to use as a hostage to defend itself. When a dominant male is displaced by a new immigrant, the new dominant will sometimes kill infants sired in the group before he arrived. Such infanticides increase the new dominant's reproductive success because lactation suppresses ovulation, and killing infants causes their mothers to resume ovulation. Other species would tell a similar story; living primates are good at discriminating between relatives and nonrelatives and behave differently toward each. It is hard to see why early hominids should have been less discriminating in their behavior.

Evidence from human behavior also challenges the big mistake hypothesis. Humans generally know who their real relatives are. No doubt propensities to cooperate with kin are deeply ingrained in human psychology. One excellent body of evidence comes from the seemingly tangential literature on incest avoidance. Westermarck (1894) suggested that there is an innate avoidance of inbreeding. If so, humans must have an innate kin recognition system. The operation of this device is nicely illustrated by the rarity of marriage among Israeli kibbutz age mates (Shepher 1983) and the poor success of Taiwanese minor marriages (Wolf 1970). In these famous examples, potential husbands and wives are raised in close companionship as children, much as siblings normally are, even though they are unrelated in fact. Coresident age mates apparently have an innate algorithm that invokes a mating avoidance mechanism. The kin-recognition system fails in the rather unusual circumstances discovered by Shepher and by Wolf, but in normal families it will function properly as an incest avoidance mechanism. If it functions for mate choice, it also ought to function to regulate cooperation. Indeed, much human cooperation is nepotistic, and nepotistic motivations often conflict with larger-scale loyalties.

Manipulation Hypotheses

When confronted with an example of what seems like cooperation among nonrelatives, some sociobiologists argue that the apparent behaviors are really the result of coercion by others. Consider the example of delayed marriage in age-graded societies in Africa, such as the Maasai of Kenya and Tanzania. Among the Maasai, every few years all of the boys near puberty are circumcised. All adult men who are circumcised together belong to an "age set." Immediately after circumcision, a man and all the members of his age set spend a period wandering throughout the countryside in an attempt to prove their bravery, for example by killing a lion on foot armed with only a spear. After several years, the members of the age set graduate to become *morani* or warriors. *Morani* provided the manpower for the successful Maasai military machine that was conquering most of East Africa when the British arrived at the end of the nineteenth century. Then after 8 or 10 years, the age set graduates and its members become older men who do not fight and who are eligible to marry. At first blush, the behavior of the *morani* seems to be an example of altruism toward unrelated individuals. Young, vigorous men seem to postpone marriage in order to benefit all the members of their tribe, most of whom are only distantly related, if at all.

Harpending et al. (1987) argue that this is not an example of altruism because the *morani* are forced to fight and to postpone reproduction by the older men in *their* own interest. While the older men may not be as powerful fighters as the younger men, according to Harpending et al., their ownership of resources and their networks of political alliances allow them to dominate the younger men. Older men use their power to force the *morani* to do the fighting and they keep the women for themselves. Thus there is no altruism to explain.

However, this explanation is at best incomplete and at worst incoherent. Let us suppose that young men obey older ones because if they do not, the older men will punish them. This fact explains why the young men fight and refrain from marriage, but it is not a complete explanation because it does not explain why older men coerce. Coercion is costly to the individuals doing the coercing. If a young man takes a wife, then some of the graybeards will have to lean on him, and this will consume lots of time, effort, and political capital (as anybody who has had to administer academic discipline knows). Moreover, this is likely to make the young man and his relatives angry, and motivate them to seek revenge. Warriors in their prime should normally be able to beat graybeards if it comes to physical coercion. Thus, the older men must receive some benefit to compensate them for these costs. Some sociobiologists seem to argue that the fact that the *morani* defend the old men and that their delayed marriage allows the old men to marry the young women is sufficient compensation. However, the fact that the older men benefit as a class is irrelevant to individual advantage. All that matters is the individual benefit that results from an older man's own coercive acts. He must marry a woman, or benefit himself directly in some other way.

Moreover, he will be indifferent about marriages that do not affect him directly. In fact, this explanation is just a different group functional explanation that substitutes benefit to a class for benefit to the whole group.

A manipulation argument is cogent if one argues that each older man enforces late marriage on the few young men with whom he interacts. But then it is incomplete because it does not explain why all old men should have such a homogeneity of interests. It is much more likely that the interests of old men are heterogeneous. Some would benefit by denying marriage opportunities to their younger colleagues, but some would not benefit because they have as many wives as they can afford, or because they want to acquire a rich or well-connected husband for their daughter, or because their own sons want a wife. An individual advantage argument predicts a variety of outcomes that depend on the details of local conditions.

Sometimes there are good reasons for homogeneity. Consider why in most hunter-gatherer societies no women hunt. The usual argument is that women don't hunt because hunting is incompatible with child care. However, this argument does not apply to some women, for example, those who are sterile. Why don't they hunt? Hillard Kaplan argues that the reason is that learning complex subsistence skills such as gathering or hunting requires many years of practice. Young boys invest in learning to hunt and young girls learn to gather because, on average, hunting yields a higher return for men and gathering for women. By the time a woman discovers she is sterile, it is too late to acquire the necessary skill to be a successful hunter. It may be possible to rescue the manipulation hypotheses with analogous arguments; however, this still remains to be done.

The manipulation hypothesis exemplifies what might be called the naive cynicism of some sociobiological reasoning. Sociobiologists reject the theoretically unmotivated group functionalism that bedevils much of social science and seek to explain seemingly altruistic behavior in terms of individual advantage. While we do favor alternative hypotheses to that of individual advantage, we also think such cynicism about human motives is an appropriate starting point, a good null hypothesis. However, some sociobiologists seem to accept any cynical explanation as if it were based on individual advantage, whereas many if not most cynical explanations of human behavior envision one group exploiting another. The fact that the older men benefit as a class is irrelevant to individual advantage; it is a group advantage. To reject the argument that young men defer marriage to benefit the group in favor of the argument that old men coerce younger ones in their own group's interest is to substitute one group functional argument for another. In a similar vein, sociobiologists have argued that restrictions on women's action in pastoral societies exist because they are in the men's interest (Strassman 1991), and that commoners have extensive restrictions on incest because it is in the interest of the aristocracy (Thornhill 1990). These explanations sound like individual advantage arguments because they invoke a conflict of interest, but neither of them really is. They each invoke group advantage—the advantage to men as a group and to the aristocracy as a group.

As we will see in the next section, it is possible to retain the manipulation hypothesis, but only if it is stripped of any sense of exploitation. We will see that the mechanism that makes manipulation possible is reciprocity, and in its simplest formulation people in essence manipulate themselves in their own group's interest.

Moralistic Reciprocity Hypotheses

A number of authors have suggested that human cooperation is based on reciprocity (e.g., Trivers 1971; Wilson 1975; Alexander 1987; Binmore 1994), and that our more sophisticated mental skills allow us to manage larger social networks than other creatures. Two kinds of evidence support this hypothesis. First, reciprocity clearly plays an important role in contemporary human societies all over the world. Second, some measures of brain size are correlated with social complexity; animal species that have small social networks tend to have smaller brains (corrected for body size) than species with large social networks (Dunbar 1992). The fact that humans have very large brains for their body size suggests that humans can maintain reciprocal relationships in larger groups than other animals. Field and laboratory experiments suggest that monkeys are much smarter about social problems than nonsocial problems. For example, vervet monkeys do not seem to know that python tracks (which are obvious and unmistakable) predict the presence of pythons, but they do know that their aggression toward another vervet predicts aggression by that individual's relatives toward them (Cheney and Seyfarth 1990), which suggests that solving social problems is important for evolution of the brain.

The defining feature of reciprocity is that ongoing interactions allow people to monitor each other's behavior and thereby reward cooperators and punish noncooperators. Beyond this property, there is little agreement among biologists or anthropologists about the details of how reciprocity works. In the simplest models, punishment takes the form of withdrawal of further cooperation (for example, Axelrod and Hamilton 1981): "I will keep helping you as long as you keep helping me, but if you cheat, I won't help you any more." We will refer to such strategies as simple reciprocity. Other authors (e.g., Binmore 1994) argue that punishment takes other forms; noncooperators are punished by various forms of social ostracism, reduced status, fewer friends, and fewer mating opportunities. Following Trivers (1971) we will call this moralistic reciprocity. While these different types of reciprocity are often lumped together, they have different evolutionary properties.

It is unlikely that large-scale human cooperation is supported by simple reciprocity. There is strong theoretical support for the idea that lengthy interactions between pairs of individuals are likely to lead to the evolution of this kind of reciprocating strategy (see Axelrod and Dion 1988; Nowak and Sigmund 1993 for review), but recent work suggests that simple reciprocity cannot support cooperation in larger groups (Boyd and Richerson 1988, 1989). An increasing group size places simple reciprocating strategies on the horns of a dilemma. Reciprocating strategies that tolerate a substantial number of defectors in the group and allow defectors to go unpunished cannot persist when common because defectors obtain the benefits of long-term cooperation without paying the cost. Thus, reciprocators must be provoked to defect by the presence of even a few defectors. However rare, intolerant strategies cannot increase unless there is a substantial chance that groups made up mainly of reciprocators will form when reciprocators are rare. This dilemma is not serious when pairs of individuals interact; minor perturbations allow rate reciprocating strategies to increase in frequency. As interesting groups become larger, however, both of these requirements become impossible to satisfy.

This conclusion makes intuitive sense. We know from everyday experience that reciprocity plays an important role in friendship, marriage, and other dyadic relationships. We stop inviting friends over to dinner if they never reciprocate; we become annoyed at our spouse if he or she does not take a turn watching the children; and we refuse to return to an auto repair shop if they do a bad job. However, it is not plausible that each of a thousand union members will stay out on strike because they are afraid that their defection will break the strike. Nor does each member of a Mae Enga war party in Papua New Guinea maintain his position in the line of battle because he fears that his desertion will precipitate wholesale retreat.

Moralistic reciprocity provides a much more plausible mechanism for the maintenance of large-scale cooperation. Reciprocators can punish noncooperators in many ways besides withholding their own cooperation. Strikebreakers can be physically attacked or their property can be vandalized. Even more plausibly, they can be socially ostracized; scabs lose status in their community and with it, many important benefits of social life. Much the same goes for cowards and deserters, who may be attacked by their erstwhile compatriots and shunned by their society, made the targets of gossip, or denied access to territories or mates.

Moralistic reciprocity enforced by such punishment is more effective in supporting large-scale cooperation than simple reciprocity for two reasons: When a simple reciprocator stops cooperating to punish defectors, he or she induces other reciprocators to stop cooperating. This, in turn, induces still more defections. Innocent cooperators are in effect punished as much as guilty defectors when the only response to defection is to stop cooperating. In contrast, other forms of retribution can be targeted so that only defectors are affected. This means that defectors can be penalized without generating a cascade of defection. Second, with simple reciprocity, the severity of the sanction is limited by an individual's effect on the whole group, which becomes diluted as group size increases. Moralistic sanctions can be much more costly to defectors and therefore can allow rare cooperators to induce others to cooperate in large groups.

However, there is also a problem with moralistic reciprocity that remains to be explained: Why should individuals punish? Remember the problem with the manipulation hypothesis: If punishing is costly and the benefits of cooperation flow to the group as a whole, administering punishment is a costly group-beneficial act, and therefore selfish individuals will cooperate but will not punish. The elder Maasai who punishes a younger man suffers a cost to himself and provides a benefit to his fellow oldsters. Similarly, the striker who attacks a scab may be injured himself. The striker who shuns a scab may forgo a satisfying friendship, a beneficial business relationship, or even a desirable marriage partner. Thus, as long as the effect of the punishment administered by a single individual will have little effect on the success of the strike, selfish individuals will not punish.

This problem is solved if moralistic reciprocators also punish people who do not punish when they should. This means that moralistic strategies punish noncooperators, individuals who do not punish noncooperators, and individuals who do not punish nonpunishers. When such strategies are common, rare noncooperators suffer because they are punished. Individuals who cooperate but do not punish suffer because they are also punished when they do not punish. In this way, it can pay to punish, even though the cooperation that results is not sufficient to compensate individual punishers for the cost of punishing. The Maasai graybeard who arranges for his son to marry will lose status in the community; the striker who fails to shun the scab will be shunned; and the Enga warrior who fails to participate in punishing a coward may become a "garbage man" himself. If moralists are common, the cost of punishing rare noncooperators may be small and the cost of being punished may be large, so that even quite costly group-beneficial behaviors can be maintained by this mechanism. [See Boyd and Richerson (1992) for a more formal version of this argument. There is also a closely analogous result in economics; see Hirshleifer and Rasmusen (1989) or Binmore (1994).]

There is also a big problem with the moralistic reciprocity hypothesis. It explains how costly group-beneficial behavior can persist, but it provides no explanation for why group-oriented behavior is more common than any arbitrary behavior. Moralistic punishment can stabilize any arbitrary behavior, e.g., wearing a tie, being kind to animals, or eating the brains of dead relatives. It does not matter whether the behavior produces group benefits. All that matters is that when moralistic punishers are common, it is more costly to be punished than to perform the required behavior, whatever it might be. When any behavior can persist at a stable equilibrium, then the fact that reciprocity is at a stable equilibrium does not tell us whether or not it is a likely outcome.

What we need to know is, does the fact that reciprocity leads to beneficial outcomes for everyone make it more likely that moralistic punishment will support reciprocity than, say, eating the brains of dead relatives. If the answer to this question is "no," then moralistic reciprocity is not a complete answer to the puzzle of human cooperation. To answer this question, we need to know how a population changes through time. Our approach is to model behavioral change as a process of cultural evolution by biased transmission. People differ in their beliefs and values about what behaviors should be subject to moralistic punishment. Depending on the composition of the population, some beliefs will yield higher payoffs than others, and therefore people with those beliefs will be imitated and they will spread. This process will then go on until the population reaches a stable equilibrium. We have analyzed a model of the evolution of moralistic punishment that incorporates these ideas, and it suggests that the existence of a group benefit does not increase the ability of a strategy to spread when it is rare (Boyd and Richerson 1992). Cultural evolution by biased transmission is equally likely to reach an equilibrium at which people are punished for not eating the brains of dead relatives as an equilibrium at which they are punished for not defending their group.

One answer to this objection (Binmore 1994) is that people don't have to slowly evolve toward a beneficial state through a myopic process of cultural evolution. They can deliberately choose what behaviors should be punished, either by getting together and talking about it, or by delegating their choice to a king, council of elders, or some other authority. There is little doubt that collective choice may account for some cases, but it seems to us that there are many more cases that it cannot account for. Think about wearing a tie. In the business world you will surely be punished if you do not wear a tie, yet this convention was never deliberately chosen. Or think about Nuer marriage practices. At the time of the Nuer expansion, there were more than 100,000 Nuer living in twelve politically independent and often hostile tribes. Their marriage rules were not chosen; the cultural group was much, much larger than any social or political entity that could conceivably do any choosing. Outside modern industrialized nation-states, it is quite common for the scale of social interaction and political organization to be much smaller than the scale of cultural similarity, and in such cases we do not see how deliberation can work. Furthermore, deliberation, when it works, is most likely founded on a basis of trust in kings, councils, or voters to create rules in the common interest. At least, modern democratic systems seem to work much better when they are founded on popular communal institutions and work poorly in communities where such civic institutions are weak (Putnam 1993). Suspicious, rational, selfish punishers may in principle deliberately choose rules to be mutually enforced, but we suspect that such a regime will prove empirically if also not theoretically fragile.

Cultural Group Selection

The idea that there is selection among societies has a certain empirical plausibility. The Nuer and the Dinka provide the model. According to Kelly (1985), differences in marriage practices led to the Nuer military superiority over the Dinka. The Nuer have a higher and less flexible bride price than the Dinka, requiring households to maintain close relations with a larger group of allied households so as to be able to assemble the requisite number of cows. The Nuer do not seem to have deliberately invented the practice of requiring a larger bride price to gain a military advantage, but the larger web of social connections maintained for bride price reasons allows them to assemble considerably larger fighting forces than the Dinka and thus provides a persistent military advantage. If we accept Kelly's analysis, then it follows that selection between the Nuer and the Dinka favored Nuer marriage practices.

Cultural group selection requires that:

1. There must be differences among groups.
2. These differences must affect the persistence or proliferation of groups.
3. These differences must be transmitted through time.

If these three conditions hold, then all other things being equal, cultural attributes that enhance the persistence or proliferation of social groups will tend to spread. The Nuer and the Dinka clearly differed; this difference allowed the Nuer to expand at the expense of the Dinka and new Nuer groups continued to behave like Nuers even though they occupied lands previously held by the Dinka. Although the Nuer example is especially unambiguous, it is not unique. There are many examples of human groups who disappeared because they could not cope with the environment or other human groups, and it is plausible that in many cases some groups survived while others did not because of cultural differences.

Why Group Selection Normally Does Not Work

The big problem with cultural group selection is theoretical. How can cultural variation among groups be maintained against the corrosive effect of self-interest? To understand this problem, consider the following simple model: Suppose that individuals live in groups embedded in a larger population. The groups represent the "societies" whose extinction drives cultural group selection. Suppose, for simplicity, that each individual in the population is characterized by one of two cultural variants. One variant causes individuals to place a high value on group goals compared with personal gain. Under the right circumstances, these "cooperators" will act in the groups interest even if it is personally costly. The other cultural variant causes individuals to place a low value on group goals compared with personal gain. In the same circumstances, these "noncooperators" will not cooperate. These cultural variants evolve under the influence of four evolutionary forces:

Biased cultural transmission People observe the consequences of the behavior of others. If, as we assume, cooperation is costly to the individual, most people will see that noncooperators do not suffer as a result of their noncooperation, and will cease their costly cooperation. However, a smaller number will err and conclude (incorrectly) that cooperation is individually beneficial. Most of the time, biased cultural transmission will cause the proportion of cooperators in each group to decrease.

Random cultural drift   If subpopulations are small, there may be random changes that are due to sampling variation. For example, sometimes by chance most people will erroneously conclude that cooperation does pay and during that generation, the frequency of cooperation will increase. This effect will be particularly important if a few prominent people play a disproportionate role in cultural transmission.

Migration among groups   No society is completely isolated from other societies. People leave and are replaced by immigrants who bring with them different culturally acquired beliefs and values. People also borrow ideas from people in other societies. Both kinds of mixing cause the proportion of cooperators in each subpopulation to approach the average proportion in the population as a whole. In other words, migration will cause groups to become more alike.

Selection among groups   The probability that a group will become extinct is negatively related to the proportion of cooperators—the more cooperators, the lower the probability of extinction. When a group becomes extinct, its territory is occupied by people from surviving groups.

Each of these forces has different effects. Driven by self-interest, biased transmission causes noncooperative behavior to increase within every group. Group selection works in the opposite direction, increasing cooperative behavior. Drift and migration affect the pattern of cultural variation within and among groups. We want to know under what conditions will group selection cause the group-beneficial behavior to come to dominate the population?

The answer is: "hardly ever." The model we have just sketched is closely analogous to interdemic group selection models studied by population geneticists. The only real difference is that biased transmission rather than natural selection leads to the spread of selfish behavior within subpopulations. Extensive analysis of such models by population geneticists indicates that populations will eventually become composed almost completely of noncooperators unless subpopulations are extremely small, there is little mixing among groups, and it is very difficult for individuals to discern the costs and benefits of alternative beliefs. Selection among groups fails because biased transmission and migration combine to create a powerful force that reduces variation among groups. Migration among groups ensures that noncooperators are always present in every group, and biased transmission causes their frequency to increase.

In contrast, the only force that generates differences among groups—random fluctuations caused by small population size and small numbers of individuals colonizing empty habitats—is weak unless groups are very small. Thus there will be little variation among groups, and even if group selection is very strong, group selection can have little effect. The existence of cultural transmission is by itself no guarantee that there will be enough variation among groups to allow cooperation to evolve.

The impatient anthropologist might say: "Who cares. We anthropologists know that there is heritable variation among societies, so group selection will work. If sociobiologists and economists and like-minded folks need an account based on individual decisions, let them worry about it. We should get on with the business of determining the function of cultural practices." Obviously we think that it is important to reconcile anthropological and evolutionary explanations, but many anthropologists do not. Even so, it is a serious mistake to ignore the problem of the maintenance of variation for cooperation between societies. All societies have internal conflict, and the balance between cooperation and selfishness varies considerably from one society to the next. The basic empirical facts point to a dynamic balance between selfish and prosocial motives and much complex structure regarding what behavior is appropriate toward comembers of different groups (e.g., comembers of one's family, extended family, ethnic group, religion, and nation). Humans are not unconditional cooperators, so we cannot assume that any given social institution is group functional even if many do have such elements. The cases are quite evidently diverse and the devil is in the details of the evolutionary processes in operation (for a nice comparative study, see Knauft 1993). In the remainder of this chapter we outline two mechanisms that maintain variation in costly group-beneficial behavior. Each mechanism is consistent with human behavior having been shaped by natural selection. However, each mechanism also has empirical requirements that need to be tested with anthropological data; each mechanism makes additional predictions; and each mechanism points to interesting new questions about the logic of the group selection process.

How Group Selection Can Work

For group selection to be an important process that can generate adaptations at the group level, there must be some mechanism that can maintain variation among groups, and not just for some memes, but for any arbitrary meme. We think that there are at least two such mechanisms: moralistic reciprocity and frequency-dependent bias. Let's see how they work.

Variation Is Maintained by Moralistic Reciprocity   Group selection can be an important process, generating group-beneficial behavior when such behavior is enforced by moralistic punishment. Moralistic punishment allows a wide variety of individually costly memes to persist once they become common, and as a result moralistic punishment provides a mechanism for preserving variation among groups, variation that then can be subjected to selection at the group level. By itself, moralistic punishment does not provide an explanation for human cooperation, but the combination of moralistic punishment and group selection does.

To show how moralistic reciprocity allows group selection to work, we modify the group selection model described earlier. Once again a population is subdivided into a fixed number of groups linked by migration, and there are two memes. However, now each meme specifies a different set of behaviors that are to be enforced by moralistic punishment. One specifies that a person's primary loyalty is to his or her kin, and people who violate this norm are punished. The second specifies that a person's primary loyalty is to his or her group and, again, violators are punished. We will call these the nepotistic and the nationalistic memes, respectively. All other things being equal, people are biased in favor of nepotism, but when the meme for nationalistic group loyalty is sufficiently common, the effects of punishment overcome this bias and people tend to choose the nationalistic meme. Until recently, everybody in the population was nepotistic, but then most people in one group adopted the nationalistic meme. We will return to the question of why this should occur, but for the moment we simply regard it as a random event, the group-level analogue of a mutation. The mutant group has a competitive advantage because nationalistic beliefs allow the organization of larger and more inclusive corporate activities, and as a result it is more likely to survive as a group and more likely to grow in size. When a group becomes extinct, its territory is occupied by others. The question is, under what circumstances will this group-level advantage cause nationalistic beliefs common in the mutant group to spread throughout the population?

First, biased transmission must be strong enough to keep the nationalistic meme common in the mutant group. To see why, suppose this meme is rarely expressed and rarely punished, so it is difficult for immigrants to learn that nepotism will get them in trouble, and therefore biased transmission is weak. As a result, only a few immigrants will have switched to nationalism by the time more immigrants arrive. This reduces the penalty for being nepotistic, and thus even fewer of the new immigrants will switch during the next time period. Eventually nepotism becomes advantageous, and the initially nationalistic group becomes nepotistic like the population as a whole. Unless biased transmission is strong enough to resist the homogenizing effects of migration, nationalism cannot spread by group selection because it cannot persist long enough for selection among groups to occur.

In contrast, suppose people with nepotistic beliefs are frequently punished in the nationalistic group so that new immigrants whose beliefs differ from those of the majority rapidly learn that such beliefs get them in trouble and they adopt the prevailing norm. Then when yet more nepotistic immigrants arrive, they find themselves to be in the minority. They quickly learn the local norms and so maintain the nationalistic meme, which originally arose merely by chance, at a high frequency long enough for group selection to have some effect.

Second, group reproduction must preserve cultural differences. Suppose that one of the groups in which nepotistic memes dominate becomes extinct. If their territory is occupied by individuals drawn from all other groups, the majority of the colonists will be nepotistic and as a result the new group will be nepotistic. Even if the group in which the nationalistic meme is common never becomes extinct, nationalism will not spread because the nationalistic group cannot reproduce itself. In contrast, suppose that new groups are formed by the fissioning of existing groups. Then most of the time when a nepotistic group fails, it will be replaced by a new nepotistic group, but occasionally it will be the nationalistic group that fissions, and when this occurs the nationalistic meme spreads. If nationalistic groups are more likely to survive or more likely to grow and fission, the result will be that nationalism will spread.

The fact that this mechanism only works when biased transmission is strong explains why it is likely to be more important for cultural evolution than for genetic evolution and therefore provides a potential explanation for why humans are more cooperative than other animals. Evolutionary biologists normally think of selection as being weak, and although there are many exceptions to this rule, it is a useful generalization. For example, if one genotype had a 5 percent selection advantage over the alternative genotype, this would be thought to be strong selection. Suppose that a novel group-beneficial genotype has arisen and that it has become common in one local group, where it has a 5 percent advantage over the genotype that predominates in the population as a whole. For group selection to be important, the novel type must remain common long enough to spread by group selection, which is possible only if the migration rate per generation is less than 5 percent (see Boyd and Richerson 1990 for details.) This is not very much migration. The migration rate between neighboring nonhuman primate groups is on the order of 50 percent per generation. While migration rates are notoriously difficult to measure, it seems likely that these rates are typically high among small local groups that suffer frequent extinction. Migration rates between larger, deme-sized groups are much lower, but so will be the extinction rate. If one assumes that migration rates are the same order of magnitude for cultural evolution, then it seems likely that biased transmission can easily maintain intergroup variation.

Variation Is Maintained by Conformist Social Learning   Frequency-dependent bias can also maintain variation among groups. We have shown (Boyd and Richerson 1985; Henrich and Boyd 1998) that natural selection can favor a psychological propensity to imitate the common type. When it is difficult but not impossible to determine what is the best way to behave in a local environment, you may increase your chances of doing the right thing by doing what everybody else is doing. The propensity to imitate the common type creates an evolutionary force that causes common memes to become more common and rare memes to become more rare. If this effect is stronger than the effect of direct bias, variation among groups will be maintained.

Once again consider our basic group selection model. As before, there are a number of groups linked by migration. Now, however, assume that the two memes affect religious beliefs. "Believers" are convinced that moral people are rewarded after death while the wicked suffer horrible punishment for eternity, whereas "nonbelievers" do not believe in any afterlife. Because they fear the consequences, believers behave better than nonbelievers; they are more honest, charitable, and selfless, and as a result, groups in which believers are common are more successful than groups in which nonbelievers are common. People's decision to adopt one meme or the other is only weakly affected by direct bias. People seek comfort, pleasure, and leisure, and this causes them to prefer to behave wickedly, at least on occasion. However, it also causes them to prefer to avoid the endless searing pain of an eternity of burial in a burning tomb. Since people are uncertain about the existence of an afterlife, they are not strongly biased in favor of one meme or the other. As a result, they are strongly influenced by the meme that is common in their society. People who grow up surrounded by believers choose to believe, while those who grow up among worldly atheists do not.

The group-beneficial belief in an afterlife will spread if two conditions are satisfied. Frequency-dependent bias must be strong enough to overcome the effect of migration, and new groups have to be formed through the fissioning of existing groups. These conditions are similar to those under which moralistic reciprocity allows the spread of group-beneficial traits. The reason is that both processes lead to bias against rare memes—moralistic reciprocity because carriers of rare memes are punished and frequency-dependent bias because, averaged over many traits, rareness indicates undesirability. In either case, when the bias against rare types is strong compared with mixing, biased transmission can maintain differences among groups. However, the same bias against rare types means that successful beliefs can spread only if groups reproduce themselves as coherent entities so that successful memes remain common.

The fact that variation is maintained by bias against rare memes also determines what is a group extinction. In genetic models, group extinction requires the death of group members. However, the "extinction" of a cultural group only requires the disruption of the group as a social unit and the dispersal of its members. As long as social extinctions are not too common, such dispersal has much the same effect as physical extinction because the dispersing individuals will constitute a small fraction of any group they join. In most cases they will temporarily change the frequency of the alternative meme in their new group a small amount, after which it will return to its original composition (which is determined by the balance of migration and the bias against rare types).

The form of group selection outlined here can be a potent force even if groups are usually very large. For a group-beneficial meme to spread, it must become common in an initial subpopulation. The rate at which this will occur through random drift-like processes (Cavalli-Sforza and Feldman 1981) will be slow for sizable groups (Lande 1985). However, this need occur only once. Thus, even if groups are usually large, occasional bottlenecks in some groups could allow group selection to get started. Similarly, environmental variation in even a few subpopulations could provide the initial impetus for group selection.

The difference between moralistic reciprocity and biased social learning is illustrated by the different answers they give to the question, why do most people growing up in a devout Christian society end up believing in the tenets of the Christian faith? People who believe that cultural variation is maintained by moralistic punishment would explain this difference in terms of reward and punishment. People who do not adopt Christian beliefs in a devout Christian society are punished, and people who do not punish such heretics (say, by continuing to associate with them) are themselves punished. People adopt the prevalent belief because they calculate (not necessarily consciously) that it yields the highest payoff. People who believe that cultural variation is maintained by conformist transmission and similar cultural mechanisms would view young people adopting the tenets of Christianity as accurate descriptions of the world (God exists; He intervenes in human affairs, and so on) because such beliefs are widely held and difficult for individuals to prove or disprove.

Each mechanism has limitations. Moralistic reciprocity can only explain adherence to behaviors that are enforced by punishment. We think that there are many examples of people adhering to costly norms even though there is little risk of punishment, and we think most readers do too—even those who think they don't. Here is a self-test. Do you believe anything that you read in ethnographies? If you do, then you agree with us. Anthropologists go off by themselves to study some exotic culture, and after spending a few years observing and talking to the locals, they return and write up their results. Ethnographers have a large incentive to come back with interesting or impressive results—their careers depend on it—and thus they have a strong incentive to fudge or make up their data. What's more, they could cheat with little risk of being caught or punished. Their observations are rarely replicated, and even when they are, the world is noisy and changeable. Thus, if you think that people obey norms only when they fear punishment, you should distrust all ethnography. While we think that the ethnographic method is far from perfect, we also think that ethnographers for the most part do their best to accurately report their experiences.

Frequency-dependent bias can potentiate group selection only if it is stronger than direct bias, and this can occur only if it is difficult for individuals to evaluate the costs and benefits of alternative memes. Clearly there are many group-beneficial beliefs whose costs of violation are all too obvious; e.g., should you cheat on your taxes or fake illness to avoid military service? To explain the evolution of honest taxpayers or patriotic volunteers, you perhaps need only to invoke norms enforced by punishment. However, there are also many such memes whose effects are hard to judge. Will children turn out better if they are sternly disciplined or lovingly indulged? Is smoking marijuana harmful to one's health? Is academic life a promising career in 2005? These are difficult questions to answer, even with all of the information-gathering and processing resources of a twenty-first-century industrial state. For most people at most times and most places, even questions like "Does drinking dirty water cause disease?" and "Can people affect the weather by appeals to the supernatural?" are difficult to answer. While it may be difficult to determine which meme is best, the choice may nonetheless have a profound effect on people's behavior, including behavior driven by rational choice.

Is the Model Empirically Plausible?

Knowing that cultural group selection is logically possible is useful. Some authors have claimed that social and cultural institutions cannot be understood as benefiting the group because this is inconsistent with a theory built from the actions of selfish individuals.

Given a showing of logical possibility, the interesting question is the empirical one: Have human cultures actually been shaped by cultural group selection? The models suggest that cultural group selection can lead to the evolution of group-beneficial traits if:

1. Group disruption and dispersal is common.
2. New groups are usually formed by the fissioning of existing groups.
3. There is variation among groups, and this variation affects the ability of a group to survive and produce daughter groups.

If these conditions held true for substantial stretches of human history, then it is quite plausible that human cultures were shaped by cultural group selection.

To address these questions, we (in collaboration with our colleague Joseph Soltis) have collected the relevant data from the ethnographic literature on Irian Jaya and Papua New Guinea. Until recently, most people in New Guinea lived in small, unstratified societies numbering a few hundred to a few thousand members. Such societies are characteristic of more of human history than the complex, stratified societies that dominate the contemporary world. The highlands of New Guinea were not explored by Europeans until the 1930s and were not subjugated by Europeans until after the World War II. As a result, we have high-quality ethnographic data describing peoples whose societies had not been drastically altered by a colonial administration. We read as many ethnographies as possible, looking for those with accounts of group extinction, new group formation, and variation among local groups (Soltis et al. 1995).

The data from New Guinea indicate that group-level extinctions are quite common. You need three numbers to estimate extinction rates: the number of groups in the population, the number of extinctions, and the time period during which the extinctions occurred. We were able to find these data for five cultural areas in New Guinea,

Table 5.1

Group extinctions in Papua New Guinea

Region	Number of groups	Number of extinctions	Number of years	% groups extinct every 25 years	Source
Mae Enga	14	5	50	17.9	Meggitt (1977)
Maring	13	1	25	7.7	Vayda (1971)
Mendi	9	3	50	16.6	Ryan (1959)
Fore/Usurufa	8–24	1	10	31.2–10.4	Berndt (1962)
Tor	26	4	40	9.6	Oosterwal (1961)

and all five reveal high rates of group extinction. The percentage of groups estimated to suffer extinction in each generation (25 years) ranges from about 8 percent among the Maring to about 31 percent among the Fore (see table 5.1). Moreover, more than half of the twenty-eight ethnographies surveyed mention at least one extinction, and it is likely that this underestimates the actual number because the failure to mention an extinction does not necessarily mean it did not occur.

Second, the data indicate that new groups are formed by the fissioning of existing groups. Detailed accounts for two cultures, the Mae Enga and the Mendi, report that new clans are formed when subclans grow to the size that allows them to take on the social functions usually performed by clans. Anecdotal evidence from other ethnographies indicates that the same general processes operate in other parts of New Guinea and Irian Jaya as well. We could not find any account in which colonists of new land were drawn from multiple groups, so the group reproduction process would favor the preservation of between-group variation.

Finally, the data also support the existence of cultural variation among local groups, although there is little evidence connecting this variation to the survival of local groups. We had difficulty finding good ethnographic descriptions of cultural variation among local groups, but for three regions, there is documented local variation. The Ok communities in the fringe highlands vary in terms of ritual and social organization. Tor tribes vary in language and kinship terminology. Abelam groups differentially accepted and rejected nontraditional religions. Furthermore, Barth (1971) argued that group selection has occurred among Ok communities. Ritual variation produced differing levels of centralized organization, and the more centralized groups were able to spread at the expense of the less centralized. At the very least, we could not disprove the hypothesis that cultural variation exists among local groups within larger cultural regions. More generally, if ethnographers are accurate reporters, consequential sociopolitical differences among different small-scale societies are ubiquitous.

Table 5.2
Time in generations required for replacement and cultural variants

Initial fraction favorable trait	Final fraction favorable trait	group extinction rate per 25 year			
		1.6%	10.4%	17.9%	31%
0.1	0.9	192	40.0	22.3	11.8
0.01	0.99	570	83.7	46.6	24.8

Rates of Change

The New Guinea data on extinction rates allow us to estimate the maximum rate of cultural change that can result from cultural group selection. For a given group extinction rate, the rate of cultural change depends on the fraction of group extinctions that are the result of heritable cultural differences among groups. If most extinctions are due to nonheritable environmental differences (e.g., some groups have poor land) or bad luck (e.g., some groups are decimated by natural disasters), then group selection will lead to relatively slow changes. If most extinctions are due to heritable differences (e.g., some groups have a more effective system of resolving internal disputes), then cultural change will be relatively rapid. The rate of cultural change will also depend on the number of different, independent cultural characteristics that affect group extinction rates. The larger the number of different attributes, the more slowly will any single attribute respond to selection among groups. By assuming that all extinctions result from a single heritable cultural difference among groups, we can calculate the maximum rate of cultural change.

Such an estimate suggests that group selection is unlikely to lead to significant cultural change in less than about 1000 years. The length of time it takes a rare cultural attribute to replace a common cultural attribute is one useful measure of the rate of cultural change. Suppose that initially a favorable trait is common in a fraction q_0 of the groups in a region. Then the number of generations t necessary for it to become common in a fraction q_t of the groups can be estimated using a simple formula derived in Soltis et al. (1995). The time necessary for different parameters is given in table 5.2. If one takes the extinction rate of the Mae Enga as typical, these results suggest that group selection could cause the replacement of one cultural variant by a second, more favorable variant, in no less than about twenty generations, or 500 years. Given that not all extinctions result from heritable cultural differences, 1000 years is a reasonable upper bound for the rate of evolution of a single character with a strong influence on group survival.

This rate of adaptation is not nearly fast enough to explain the many cases of cultural change known to have occurred on much shorter time scales. For example, the introduction of the horse to the Great Plains of North America in the 1500s led to the evolution of the culture complex of the Plains Indians in less than 300 years. If the rates of group extinction estimated for New Guinea are representative of small-scale societies, cultures like those of the Great Plains cannot be explained in group-functional terms. There has not been enough time for group selection to drive a single cultural attribute to fixation, even if that attribute had a strong effect on group survival.

This result also suggests that group selection cannot directly justify the practice of interpreting many different aspects of a culture as group beneficial. While both empirical and theoretical uncertainties prevent us from making a quantitative estimate of the rate of evolution of many different cultural attributes, it is clear that group selection will shape many traits more slowly than a single trait. If group selection can cause the substitution of a single trait on thousand-year time scales, the rate for many traits will be substantially longer. We know from linguistic and archaeological evidence that related cultural groups, such as the Nuer and the Dinka, which differ in many cultural attributes, often have diverged from a common ancestral group in the past few thousand years. Thus there has not been enough time for group selection to produce the many group-beneficial attributes that distinguish one culture from another.

The result does provide justification for interpreting some aspects of contemporary cultures in terms of their benefit to the group. The model demonstrates that under the right conditions group selection can be an important process, and the data from New Guinea suggest that these conditions are empirically realistic. The data also suggest that the rates of group extinction are high enough to cause a small number of traits with substantial effects on group welfare to evolve on time scales that characterize some aspects of cultural change. Group selection cannot explain why all the many details of Enga culture differ from the many details of Maring culture. It might explain the existence of geographically widespread practices that allow large-scale social organization in the New Guinea highlands, practices that evolved along with, and perhaps allowed, the transition from band-scale societies to the larger-scale societies that exist today.

Nor do our data exclude the possibility that cultural group selection explains the increase in the scale of sociopolitical organization in human prehistory and history. That evolution is slow even on the millennial scale. Anatomically modern humans appear in the fossil record about 190,000 years ago, yet there is no evidence for symbolically marked boundaries (perhaps indicative of a significant sociopolitical unit encompassing an "ethnic" group of some hundreds to a few thousand individuals) before about 35,000 years ago (Klein 1999). The evolution of simple states from food-producing tribal societies took about 5000 years, and the modern industrial state took another 5000. Evolutionary processes that lead to change on hundred-year time scales cannot explain such a slow change unless they are driven by some environmental factor that changes on longer time scales. In contrast, the more or less steadily progressive course of increasing sociopolitical complexity over the past few tens of thousands of years indeed is consistent with adaptation by group selection.

It follows from this picture that much contemporary cultural variation may be functionally significant. Critics of functionalism (Sahlins 1976; Hallpike 1986) have argued that the diversity of human societies is incompatible with functional explanations because societies faced with similar environments and using similar technologies exhibit radically different forms of sociopolitical organization. Hallpike (1986) argues that the extent of such differences in simpler societies indicates that the variation is not functional, but neutral, and that it cannot be the result of evolution. The alternative is that they represent variations currently being exposed to a slow process of group selection that is as yet far from equilibrium.

We still have the puzzle of the relatively rapid evolution of seemingly prosocial institutions on time scales that are much shorter than millennia. Most modern industrial societies have evolved from aristocratic and totalitarian forms of government to representative democracies over the past few centuries. Such societies seem to function better, at least under modern conditions, than their predecessors. None of these societies has become extinct. One possible explanation for such developments is that over the span of many millennia in the Pleistocene, basic human moral intuitions became adapted to living in group-selected cultures. The moralistic punishment mechanism, for example, will penalize individuals whose genes as well as whose memes are not conducive to following social rules. As Aristotle said, humans are those animals that by their nature live in political communities (for him, the tribal-scale Greek polis). Some measure of our prosocial impulses may have become innate, an idea we call Aristotle's hypothesis. Given some prosocial predispositions and some implementing cultural institutions that engender trust, people may individually or collectively tend to make decisions in the group interest. A considerable amount of evidence suggests that Aristotle's hypothesis is correct (Richerson and Boyd 2001).

Conclusion

Cooperation in human groups is a serious puzzle because we cooperate extensively with distantly related and even unrelated individuals. Hamilton's (1964) theory of inclusive fitness and Trivers's (1971) theory of reciprocal altruism predicted that cooperation could evolve only under highly restricted circumstances. At the time they advanced these ideas, many evolutionary biologists assumed that large-scale animal cooperation was common and that large-scale group selection for altruism was the explanation for it. The work inspired by these theories has resulted in a nearly total triumph for Hamilton and Trivers, although vigorous and sophisticated dissent is still heard (Sober and Wilson 1998). However, human politics is the only really good available example of cooperation on large and even very large scales among nonrelatives.

For the rest of the organic world, Hamilton's and Trivers's mechanisms seem virtually as unexceptional as the law of gravity. Indeed, human societies themselves show ample evidence of individual selfishness, nepotism, and small-scale cabals but nevertheless manage, to highly variable degrees, to provide the benefits of cooperation on considerable scales. While we do not regard the case closed by any means, we think that group selection on cultural variation together with moralistic punishment and Aristotelian social instincts are the strongest current candidates to explain how human societies can succeed in a world where strong selective forces favoring only small-scale cooperation act against their existence.

References

Aberle, D. F., Cohen, A. K., Davis, A. K., Levy, M. J., and Sutton, F. X. 1950. The functional prerequisites of a society. *Ethics* 60: 100–111.

Alexander, R. D. 1974. The evolution of social behavior. *Annual Review of Ecology and Systematics* 5: 325–383.

Alexander, R. D. 1987. *The evolution of moral systems.* New York: Aldine de Gruyter.

Axelrod, R., and Dion, D. 1988. The further evolution of cooperation. *Science* 242: 1385–1390.

Axelrod, R., and Hamilton, W. D. 1981. The evolution of cooperation. *Science* 211: 1390–1396.

Barth, F. 1971. Tribes and intertribal relations in the Fly headwaters. *Oceania* 41: 171–191.

Becker, G. 1981. *A treatise on the family.* Cambridge, Mass.: Harvard University Press.

Berndt, R. 1962. *Excess and restraint.* Chicago: University of Chicago Press.

Binmore, K. G. 1994. *Game theory and the social contract.* Cambridge, Mass.: MIT Press.

Boyd, R., and Richerson, P. J. 1985. *Culture and the evolutionary process.* Chicago: University of Chicago Press.

Boyd, R., and Richerson, P. J. 1988. The evolution of reciprocity in sizable groups. *Journal of Theoretical Biology* 132: 337–356.

Boyd, R., and Richerson, P. J. 1989. The evolution of indirect reciprocity. *Social Networks* 11: 213–236.

Boyd, R., and Richerson, P. J. 1990. Group selection among alternative evolutionarily stable strategies. *Journal of Theoretical Biology* 145: 331–342.

Boyd, R., and Richerson, P. J. 1992. Punishment allows the evolution of cooperation (or anything else) in sizable groups. *Ethology and Sociobiology* 13: 171–195.

Cavalli-Sforza, L. L., and Feldman, M. W. 1981. *Cultural transmission and evolution.* Princeton, N.J.: Princeton University Press.

Cheney D. L., and Seyfarth, R. M. 1990. *How monkeys see the world: Inside the mind of another species*. Chicago: University of Chicago Press.

Coleman, J. 1990. *Foundations of social theory*. Cambridge, Mass.: Harvard University Press.

Dunbar, R. I. M. 1992. Neocortical size as a constraint on group size in primates. *Journal of Human Evolution* 22: 469–493.

Frank, R. 1988. *Passions within reason*. New York: W. W. Norton.

Gintis, H. 2000. *Game theory evolving*. Princeton, NJ: Princeton University Press.

Hallpike, C. R. 1986. *The principles of social evolution*. Oxford: Clarendon Press.

Hamilton, W. D. 1964. Genetical evolution of social behavior I, II. *Journal of Theoretical Biology* 7: 1–52.

Hamilton, W. D. 1975. Innate social aptitudes in man: An approach from evolutionary genetics. In R. Fox (ed.), *Biosocial anthropology* (pp. 133–155). London: Malaby.

Harpending, H., Rogers, A., and Draper, P. 1987. Human sociobiology. *Yearbook of Physical Anthropology* 30: 127–150.

Henrich, J., and Boyd, R. 1998. The evolution of conformist transmission and the emergence of between-group differences. *Evolution and Human Behavior* 19: 215–242.

Hirshleifer, R., and Rasmusen, E. 1989. Cooperation in a repeated prisoner's dilemma with ostracism. *Journal of Economic Behavior and Organization* 12: 87–106.

Kelly, R. C. 1985. *The Nuer conquest: The structure and development of an expansionist system*. Ann Arbor, Mich.: University of Michigan Press.

Klein, R. G. 1999. *The human career: Human biological and cultural origins* (2nd ed.). Chicago: University of Chicago Press.

Knauft, B. M. 1993. *South Coast New Guinea cultures: History, comparison, dialectic*. Cambridge: Cambridge University Press.

Lande, R. 1985. Expected time for random genetic drift of a population between stable phenotypic states. *Proceedings of the National Academy of Sciences* of the USA 82: 7641–7645.

Malinowski, B. 1922. *Argonauts of the Western Pacific*. Prospect Heights, Ill.: Waveland Press (reissued 1984).

Meggitt, M. 1977. *Blood is their argument: Warfare among the Mae Enga tribesmen*. Mountain View, CA: Mayfield Publ. Co.

Nowak, M., and Sigmund, K. 1993. A strategy of win-stay, lose-shift that outperforms tit-for-tat in the prisoner's dilemma game. *Nature* 364: 56–58.

Oosterwal, G. 1961. *People of the Tor*. Assen (NL): Royal van Gorcum.

Putnam, R. D. 1993. *Making democracy work: Civic traditions in modern Italy.* Princeton, N.J.: Princeton University Press.

Radcliffe-Brown, A. R. 1952. *Structure and function in primitive society.* London: Cohen & West.

Rappaport, R. A. 1984. *Pigs for the ancestors.* New Haven, Conn: Yale University Press.

Richerson, P. J., and Boyd, R. 2001. The evolution of subjective commitment to groups: A tribal instincts hypothesis. In R. M. Nesse (ed.), *The evolution and the capacity for subjective commitment* (pp. 186–220). New York: Russell Sage Foundation.

Ryan, D. 1959. Clan formation in the Mendi Valley. *Oceania* 29: 257–289.

Sahlins, M. 1976. *Culture and practical reason.* Chicago: University of Chicago Press.

Shepher, J. 1983. *Incest, the biosocial view.* New York: Academic Press.

Sober, E., and Wilson, D. S. 1998. *Unto others: The evolution and psychology of unselfish behavior.* Cambridge, Mass.: Harvard University Press.

Soltis, J., Boyd, R., and Richerson, P. J. 1995. Can group functional behaviors evolve by cultural group selection? An empirical test. *Current Anthropology* 36: 473–494.

Spencer, H. 1891. *Essays: scientific, political, and speculative* (3 vols.). London: Williams and Norgate.

Strassman, B. 1991. The function of menstrual taboos among the Dogon: Defense against cuckoldry? *Human Nature* 2: 89–131.

Thornhill, N. W. 1990. Evolutionary significance of incest rules. *Ethology and Sociobiology* 11: 113–129.

Tooby, J., and Cosmides, L. 1989. Evolutionary psychology and the generation of culture, part I: Theoretical considerations. *Ethology and Sociobiology* 10: 29–49.

Trivers, R. L. 1971. The evolution of reciprocal altruism. *Quarterly Review of Biology* 46: 35–57.

Turner, J. H., and Maryanski, A. 1979. *Functionalism.* Menlo Park, Calif.: Benjamin/Cummings.

van den Berghe, P. L. 1981. *The ethnic phenomenon.* New York: Elsevier.

Vayda, A. P. 1971. Phases of the process of war and place among the Marings of New Guinea. *Oceania* 42: 1–24.

Westermarck, E. 1894. *The history of human marriage.* London: Macmillan.

Wilson, E. O. 1975. *Sociobiology.* Cambridge, Mass.: Harvard University Press.

Wolf, A. P. 1970. Childhood association and sexual attraction: A further test of the Westermarck hypothesis. *American Anthropologist* 72(3): 503–515.

Young, H. P. 1998. *Individual strategy and social structure: An evolutionary theory of institutions.* Princeton, N.J.: Princeton University Press.

6   From Typo to Thinko: When Evolution Graduated to Semantic Norms

Daniel Dennett

Natural selection edits with an eye only toward what the message says, not to what it means.
—Richard Powers, *The Gold Bug Variations*

Darwinian Perspectives on Culture

Do we need a Darwinian theory of cultural evolution? In one sense, certainly. It is obvious that there are patterns of cultural change—evolution in the neutral sense—and any theory of cultural change worth more than a moment's consideration will have to be Darwinian in the minimal sense of being consistent with the theory of evolution by natural selection of *Homo sapiens*. Our species name is well chosen, and it is culture that makes us the knowing hominid, so a minimally Darwinian theory of culture must hold that the phenotypic traits that make cumulative culture possible—mainly, language and the habits of sociality—evolved by natural selection, unaided by what I call *skyhooks*: saltations in Design Space that could not be the outcome of standard evolutionary mechanisms (Dennett 1995). This minimal Darwinism is simply the denial of the hypothesis that culture is, as it were, a miracle, a gift from God.[1] It maintains in one way or another that natural selection eventually provided the foundations for culture, which then took off and elaborated itself under some regime that explains the patterns of cultural change, but that regime need not itself be Darwinian in any interesting sense.

For instance, the standard model is an economic one. The theorist says, in effect, if Darwin will give me *Homo economicus*, a social group of rational, self-interested individuals getting and spending, saving and making and trading, I can then use the intentional stance (Dennett 1971, 1987) as the explanatory framework for describing and accounting for the patterns of cultural evolution. This economic model is used not just by economists, of course; it is tacitly presupposed by historians and anthropologists and all the other theorists who treat culture as composed of goods, possessions of the people, who husband them in various ways, wisely or foolishly. People carefully preserve their traditions of fire-lighting, house-building, speaking, counting, justice, etc. They trade cultural items as they trade other goods. And of course some cultural items (wagons, pasta, recipes for chocolate cake, etc.) are definitely goods, and so we can plot their trajectories using the tools of economics. It is clear from this perspective that highly prized cultural entities will be protected at the expense of less favored ones, and there will be a competitive market where agents both "buy" and "sell" cultural wares. If a new method of house-building or farming or a new style of music sweeps through the culture, it will be because people perceive advantages to these novelties and choose them. If Coca Cola bottles proliferate around the world, it is because more and more people prefer to buy a Coke. Advertising may fool them. Then we look to the advertisers, or those who have hired them, to find the relevant agents whose desires fix the values for our cost-benefit calculations. *Cui bono*? Who benefits? The purveyors of the goods and those they hire to help them, etc. In this way of thinking, then, the relative "replicative" power of various cultural goods—whether Coke bottles, building styles, or religious creeds—is measured in the market-place of cost–benefit calculations performed, consciously or unconsciously, by the people.

Biologists, too, make good use of the economic model, explaining the evolution (in the neutral sense) of features of the natural world by treating them as goods belonging to various members of various species: one's food, one's nest, one's burrow, one's territory, one's mates, one's time and energy. Cost–benefit analyses shed light on the husbandry engaged in by the members of the different species inhabiting some shared environment.[2] Not every "possession" is considered a good, however. The dirt and grime that accumulates on one's body, to say nothing of the accompanying flies and fleas, are of no value, or are of negative value, for instance. These hitchhikers are not normally considered as goods by biologists, except when the benefits derived from them are manifest.

These economic models of culture are consistent with Darwinism but are not more specifically Darwinian. Darwinian evolution provides organisms whose ultimate goal is self-replication, and who then track the rational if myopic trajectory to that end; the interactions of such rational agents determine which features of the shared environment will proliferate, which will be contested, which despoiled, and so forth. In these models, cultural traits, however they arise, spread as fitness enhancers, at least locally considered. Agriculture, cooking, clothing, the wheel, writing, the bow and arrow—all these cultural innovations are plausibly viewed as improvements that need not arise from gene mutation and recombination and need not be transmitted genetically. They are, one might say, infectious phenotypic features. These features, it is presumed, pass some sort of quality-control test administered by the agents themselves. They are chosen by evolved organisms and put to use, and if they didn't "pay for themselves" in a fitness boost (or at least an apparent fitness boost, myopically considered) they would soon die out, just like genetically transmitted phenotypic variations. The idea is that if innovations are randomly distributed around neutrality, the pernicious innovations will hasten the demise (or mating failure) of those who adopt them; the enhancements will do the opposite, and over the not very long haul the enhancers will proliferate. This vision allows the possibility, as relatively rare outliers, of mistakes: either good tricks abandoned by mistake or pernicious tricks persisted in thanks to some local illusion.

More ambitious models (Feldman and Cavalli-Sforza 1981; Boyd and Richerson 1985) then address the opportunities for coevolution, for interaction between the items that come to be present or dominant in the cultural marketplace and the genetically transmitted phenotypes of the persons transmitting and preserving those items. Clothes do make the man, at least to the extent of diverting selection pressures for weather-hardiness, so the cultural transmission of clothing sends ripples through the evolution of human physiology. Similarly, new practices of food gathering and preparation can reflexively change the fitness landscape for digestive capability. Lactose tolerance in adults descended from people who herded dairy animals is a well-studied case. These models are Darwinian in a more than minimal sense because they extend the perspective of population genetics, the replicator dynamics, to these nongenetically transmitted phenotypic features, exploring the effects of horizontal and oblique transmission, for instance. However, they also maintain the basic economic presupposition of rational agents: Cultural traits are adopted because they are deemed worth having, because they are supposed, rightly or wrongly, to contribute somehow to the achievement of one's more ulterior ends, whatever they may be.

When a rational agent or intentional system makes a decision about which is the best course of action, all things considered, we need to know from whose perspective this optimality is being judged. Here things begin to get messy. In nature, genes are the ultimate units of "self"-interest. That is to say, adaptations in plants and animals (and simpler life forms) are, by definition, features that further the *summum bonum* of gene replication, directly or indirectly. Are cultural innovations adaptations in this narrow, genetic sense? It is obvious that many cultural features are deemed by the populace to be beneficial, functional, adaptive, useful, life-enhancing, or enabling. It is less clear that these esteemed features play a discernible role in enhancing genetic fitness, as contrasted with, say, human happiness, the pursuit of which is curiously orthogonal to genetic fitness. One of the striking trends in human evolution, going back thousands of years, is the gradual diminution in the proportion of human effort devoted in any clearly discernible way to the achievement of the fundamental goals we share with animals: avoiding pain, hunger, and predation, and seeking comfort, security, and mating opportunities. Even if the peculiar human *desiderata* of prestige, power, wealth, beautiful surroundings, recreation, music, toys, and so forth, have discernible instrumental rationales (improving one's profile in the contest for mates, enlarging one's harem, one's territory, one's margin of error), they have more or less detached themselves from these inaugural foundations and become ends in themselves. The young man bought the guitar in order to attract young women, but now he has become a guitarist who would rather make music than love. As Cavalli-Sforza and Feldman note,

There are people determined to risk their life to reach the top of Mt. Everest, and others that spend their life accumulating money, or attempting artistic or scientific creations, or simply trying to do as little as possible. It is difficult to subsume all of these choices under a common schedule admitting no individual variation. (1981: 342)

As they put it,

Control is delegated to a system of poorly understood internal drives and rewards that direct the activity of the individual. . . . Our very inadequate knowledge of this steering system prevents us from making finer statements, but it is probably true that the system's overall activity is directed towards maximizing self satisfaction of the individual. Important complications arise because we can satisfy ourselves in many different, competing ways, many of which demand careful advance planning. (1981: 364)

Feldman and Cavalli-Sforza thus adopt the default assumption, at least in the Western world, and especially among economists, of treating the agent as a sort of punctate, Cartesian locus of well-being. What's in it for *me?* Rational *self*-interest. However, while there has to be something in the role of the self—something that defines the answer to the *cui bono* question for the decision maker under examination, there is no necessity in this default treatment, common as it is. A self-as-ultimate-beneficiary can in principle be indefinitely distributed. I can care for others, or for a larger social structure, for instance. There is nothing that restricts me to a *me* as contrasted to an *us*. I can still take my task to be looking out for number one while including, under number one, not just myself, but my family, the Chicago Bulls, Oxfam, the flourishing of mid-twentieth-century acoustic guitar finger-picking techniques—indeed anything human ingenuity can define and become attached to, making its welfare definitive of the decision maker's "happiness."

It is not obvious that any other organism strives for its own happiness or anything like it. If human happiness is our *summum bonum* (or at least a *bonum* against which we do in fact often attempt to measure costs and benefits), how did it arise? It is here that the prospect of a still more radically Darwinian theory of cultural evolution becomes attractive. Could the unique varieties of human evaluation that are so distantly and indirectly anchored to any plausible litmus test of genetic fitness be accounted for by supposing that human beings have evolved into a condition where they have become the vectors, the hosts, of a new order of symbionts—competing cultural replicators whose own fitness, defined in standard Darwinian terms of relative replicative success, has constituted a new sort of entity? An enculturated human being, according to this proposal, is as important a novelty on the evolutionary front as the eukaryotic cell was at its debut: a unification of distinct replicators into a synthetic organization with a displaced goal or *summum bonum*. It is no longer just an organism bent on self-replication, but a *person* bent on furthering the particular goals and ideals with which that person identifies. Has our guitarist unwittingly become part of a guitar's way of making another guitar? This is a tantalizingly attractive idea, but for such a perspective to anchor itself firmly in evolutionary theory, we must take seriously Dawkins's concept of a meme, and there are reasons to doubt that we should do that.

Cultural Replicators: A Central or a Peripheral Phenomenon?

In some neighborhoods, ball bearings outnumber rabbit turds; in other neighborhoods this imbalance is reversed; and in yet other neighborhoods no entities of either variety are to be found. These differential production patterns change over time, and there are reasons for them, but they are not, in the main, Darwinian reasons. Not all production is replication, and not all distribution is emigration. Variety and similarity are also found among cultural items, and the question is whether any (or many) of the reasons for patterns in changing "populations" of cultural items are Darwinian.

Dan Sperber (2000) notes that the dictionary definition of "memes" is too bland to be of much interest: "an element of culture that may be considered to be passed on by non-genetic means," while a more radical definition, more faithful to Dawkins's arresting proposal, "cultural replicators propagated through imitation," is far from obvious. Indeed, it is in need of defense against two objections. The "simplest and most serious objection" is that the copying of cultural items is in general too low fidelity to permit natural selection to get a purchase. Compare memes with viruses. Viruses travel light and carry no copying machines, so they reproduce by entering cells and tricking the cell's proprietary copying machines into making spurious copies of them instead of copying their usual and proper fare, the cell's own DNA. If memes are like viruses, as Dawkins and other would-be memeticists have claimed, it is because they reproduce by exploiting the copying machinery resident in the brains of human beings. How well does this parallel hold? How good is that machinery? Not good enough, it seems. We human beings are actually rather bad at the sorts of "mindless" copying that cells excel at.

Following Williams (1966), Sperber notes that although higher selection biases can tolerate lower fidelity, it still appears that "mutation" rates among memes would be so great that any description of the emerging patterns in terms of descent with modification, as Darwin put it, would be lip service only. "For memetics to be a reasonable research program, it should be the case that copying, and differential success in causing the multiplication of copies, over-whelmingly plays the major role in shaping all or at least most of the contents of culture" (Sperber 2000: 172). If we are not inveterate and talented copyists, we will make poor hosts for our cultural viruses, and Darwinian descent with (relatively rare) modification will seldom occur. We will need to look elsewhere to explain the patterns of culture.

Is it so clear that our copying is too low fidelity to work? Dawkins (1999) has responded to this objection with his example of the origami model of a Chinese junk, which people learn to make by following a canonical set of simple "self-normalizing" instructions, but Sperber finds this misleading, since "the normalisation of the instructions results precisely from the fact that something other than copying is taking place" (2000: 169). Sperber lays down three conditions for "true replication":

For B to be a replication of A,

1. B must be caused by A (together with background conditions).
2. B must be similar in relevant respects to A.
3. The process that generates B must obtain the information that makes B similar to A from A.

It is condition (3), Sperber claims, that is seldom met by cultural transmissions. Infectious laughter is his excellent example of a transmission event that meets (1) and (2) and fails (3), and he extends his analysis of this case by the fanciful example of ten sound recorders that trigger each other, but whose productions, in one case, do not consist of replications, but rather of recognitions, followed by re-productions. Triggered production of this sort is distinct from copying or replication in the one way that matters for Darwin, according to Sperber. It does not slavishly copy the original; instead it is inspired by the original to make another of the same sort—but without any systematic attempt or disposition to reproduce any idiosyncrasies of the original. It normalizes its production to an independent ideal, discarding or not even noticing any mutations, good or bad, in the original.

Sperber illustrates this point with another fine example, the contrast between a nonsense scribble and a five-pointed star. The nonsense scribble would degenerate quickly in any series of attempted replications because people are not good copiers of such productions, while the five-pointed star would be "copied" with high fidelity, just as Dawkins says. However, Sperber maintains, the succession of stars would not really be copies of their predecessors, since the "copyists" would normalize to the recipe for the drawing procedure, ignoring the details of the individual productions. Is Sperber looking at the right level of fineness? Dawkins's point is that a finite repertoire of such triggered productions is not just a good trick for human beings who want to heighten their transmission fidelity, it is a Good Trick discovered several billion years ago by natural selection. Sperber distinguishes copying from merely triggering the production of a similar effect, but a repertoire of such triggers, called an alphabet, is what makes high-fidelity copying possible, both in cells and in human culture.

Suppose Tommy writes the letters "SePERaTE" on the blackboard, and Billy "copies" it by writing "seperate." Is this copying or triggered reproduction? The normalization to all lower case letters shows that Billy is not slavishly copying Tommy's chalk marks but rather is being triggered to execute a series of canonical, normalized acts: *make an "s"; make an "e,"* etc. It is thanks to these letter norms that Billy can "copy" Tommy's word at all. And he does copy Tommy's spelling error, unlike Molly, who "copies" Tommy by writing "separate," responding to a higher norm, at the level of word spelling. Sally then goes a step higher, "copying" the phrase "separate butt equal"— all words in good standing in the dictionary—as "separate but equal," responding to a recognized norm at the phrase level. Can we go higher? Of course. Anybody who, when "copying" the line in a recipe "Separate three eggs and beat the yolks until they form stiff white cones," would replace "yolks" with "whites," knows enough about cooking to recognize the error and correct it. Above spelling and syntactic norms are a host of semantic norms as well.

DNA has an alphabet—the famous ACGT—and words, the three-letter codons that "spell" the twenty amino acids. In fact, the high fidelity of genetic transmission depends on the subcellular machinery being triggered to "recognize" and "re-produce" a small repertoire of types, whose idiosyncracies, if any, are ignored, not slavishly copied: *"make a cytosine," "make a guanine,"* etc. There are error-correcting enzymes as well, but they don't ascend (as far as we know!) above the level of a spell checker, correcting "typos" by brute template matching against the original.[3]

Does the human capacity (and irresistible disposition) to respond to higher, semantic norms—our capacity to correct not only typos but what hackers call *thinkos*—rule out cultural transmission as a candidate for natural selection? Sperber seems to think it does. "Contrary to what Dawkins writes," he claims, "the instructions are not 'self-normalizing.' It is the process of attribution of intentions that normalizes the implicit instructions that participants infer from what they observe" (Sperber 2000: 171). Sperber is partly right; the attribution of intentions is the key difference between this sort of human transmission and genetic replication. The point comes out even more clearly if we mutate Sperber's example slightly, adding a point to his star. Consider the fact that there are two distinct recipes (and many other less obvious ones) for making a regular six-pointed star:

1. Make a regular hexagon and put equilateral triangle points on each side.
2. Make an equilateral triangle and superimpose on it another one, upside-down.

A series of six-pointed star "replications" might be created by a random alternation between these two recipes with no loss of fidelity. Which recipe did various individual copyists follow? It wouldn't matter, because what is being copied is not the recipe but the result understood as an intended object having certain features.[4]

Sperber thinks that this reliance on attribution of intentions on the part of the copyists disqualifies cultural transmission as a Darwinian process of natural selection. He supposes that this invocation of intelligent, semantically sensitive, intention-attributing agents in the purported replication process flies in the face of a fundamental requirement of Darwinian processes: mindless, purposeless mechanicity. He is almost right. To see the force of this interesting objection, imagine a creationist variant on standard neo-Darwinian genetic evolution. It postulates that God watches over each moment of DNA replication and whenever He sees some copying that is "wrong" (relative to God's great plan), he undoes it. Thus when He chooses, he lets mutations flourish, and when He does not, those mutations get corrected by a gentle miraculous nudge of the error-correcting enzymes. Here Intelligence is playing a guiding role in evolution—just the sort of role that orthodox (devout, "fundamentalist") Darwinians abjure. As Richard Powers has observed, "Natural selection edits with an eye only toward what the message says, not to what it means" (1992: 546). Clever human beings, in contrast, edit with an eye toward meaning. If such clever editors are inserted into the process of cultural transmission and revision, what would be left of a Darwinian theory of culture?

This worry ignores the fact that *Homo sapiens* is not itself a miracle, a skyhook, but something that has evolved by nonmiraculous natural selection. Its capacity to respond to semantic norms is itself something that has evolved under a regime that could not respond to semantic norms. Before there could be eyes, good for distal perception, there had to be mere photosensitive responders to proximal stimulation, out of which eyes could gradually be built. Before there could be minds, good for semantic discrimination, there had to be copying machinery that could only discriminate alphabet letters. Put otherwise, DNA error-correcting enzymes have always responded to semantic norms, but just local or proximal semantic norms—*make a G*—as contrasted to more distal semantic norms—*make a codon for asparagine* or *make some lysozyme* or *make a protein that blocks serotonin uptake*, or even *make something that will fight off infection*.

Why shouldn't evolution go right on working once the copying machinery graduates to less myopic norms? Even our lowest-level mindless copying avails itself of correction to a norm. Is there a "highest permissible" level of normalization in any Darwinian process? Darwin (and Fisher, and Williams, and others) saw the need for a sufficiently "strong principle of inheritance" to keep evolution going, but nothing has been said about how that fidelity is to be maintained mechanically. Let there be copyists that take themselves to be responding to semantic norms; there will still be a suitably long-distance evolutionary perspective from which their copying efforts, for all their editorial work, will appear myopic and unwitting, oblivious to—and hence unresponsive to—the larger-scale pattern of differential replication that ensures that a Darwinian process is occurring.

In "Pierre Menard, Author of the *Quixote*," Jorge Luis Borges (1962) tells the fanciful tale of a literary theorist who sets out to *compose* (not copy, not write from memory) Cervantes' great work anew in the twentieth century. This will be an act of bizarre self-control, since Menard is a Cervantes scholar who no doubt has at least large portions of the text of *Don Quixote* committed to memory, but Menard is determined to bracket that memory and create, with his own authorial intentions, all of Cervantes' sentences anew, like an experienced wheelwright setting out to reinvent the wheel! He succeeds (though how can he tell?), and Borges tells us: "Cervantes' text and Menard's are verbally identical, but the second is almost infinitely richer" (1962: 42). In one sense, Menard did not copy or memorize Cervantes' text, but in another sense, he did, in spite of his virtuoso self-control, his obsessive act of re-creation. He did, because, as Sperber puts it, "(3) The process that generates B must obtain the information that makes B similar to A from A" (2000: 169) and surely Menard's prior study of Cervantes' text is an essential part of the scholarship that permits him to "compose" *Don Quixote* anew. Of course Menard has used a lot of other information as well; the surplus is presumably what permits him, unlike an ordinary reader, to claim to have re-composed, not written down from brute memory, the work. But so what? According to Borges, the texts are "verbally identical," so high-fidelity reproduction has occurred. Imagine a world in which Menards abound, devoting their lives to the re-composition of their favorite works. The transmission of texts will proceed just fine in such a world—as fine as if photocopying machines were the underlying machinery.

In fact, of course, a pastel version of that fantasy is just what has happened in the transmission of ancient texts in our world, for seldom if ever have the scribes taking dictation been entirely uncomprehending of the words they were dutifully "copying," and so they have willy-nilly "corrected" whatever they heard in the process of transcribing it. Their corrections have been governed by several levels of norms: orthographic or lexical, syntactic, and finally semantical. The imaginary Menard can be conceived to have "transcribed" the entire poem of Cervantes *modulo* the "semantic norm" of the whole text. Most of us lack that highly sophisticated norm; we tend to fall back on our sense of the gist of such a narrative, or when all else fails, rote memory or parrotting (but even "parrotting" is not like a parrot's parrotting—unless it is, as it very seldom is, a matter of reiterating formulas in a language we don't understand).

When Sperber notes that in cultural transmission "the information provided by the stimulus is complemented with information already in the system" (2000: 171) he is right, but the same is true of DNA replication. The main difference, so far as I can see, is that unlike DNA replication, human cultural replication is accomplished by processes of highly variable semantic depth, responding to perceived (and misperceived) "copying" errors relative to norms at many levels. The alphabets of written languages provide us with the most vivid and best-understood system of such norms of replication, but the phenomenon of semantic norms is not directly tied to language. Musical notation relies on the staff to digitalize the roughly inked spots, so that a musician can see at a glance that a chord is A-C#-E-G even though the A is written almost twice as far beneath the staff as it is "supposed" to be. A sketch of a new sort of axle for a wagon need not make the wheels exactly round; the user of the sketch will recognize those irregular closed curves as representations of wheels, which are to be round, of course. As we move through our various apprenticeships in life, we learn to perceive new families of categories—new alphabets, in an extended sense—from which to construct high-fidelity copies. Only a skilled potter can see at a glance what another potter is doing and copy it or teach it to others.

Consider a chef demonstrating the making of a sauce to an apprentice. The description in words might be "deglaze the pan, reduce the sauce, and thicken with cornstarch," but the words aren't really necessary if the apprentice appreciates the goal of each process. Here is a series of three analog processes, none of which could be exactly copied by the apprentice. The cook didn't measure the water he sloshed into the pan, didn't time the reduction period, and added cornstarch freehand until the sauce took on the desired consistency. The recipe can be transmitted faithfully all the same, thanks to the shared norms for these analog processes already inculcated in the apprentice.

If the error correction in the case of memes is semantically appreciative at many levels, doesn't this show that cultural evolution is *not* a mindless algorithm but rather a system that must invoke high-level semantic comprehension at every juncture? The variable depth of semantic norms does guarantee that memeticists will have a problem providing identity conditions for memes that are more severe than the (already severe) problems afflicting the identity conditions for genes. If we consider that the meme ought to be understood to be the smallest unit of *information worth copying*, then we have already accumulated a wealth of understanding of just such problems, which arise in patent law and the law of copyright and trade secrets. How big is an idea? When is one idea an illicit copy of another idea? We have no single bedrock criterion for answering such questions, but we manage quite well with them in practice, counting on the costs of re-invention to stabilize our sense of what is worth copying in particular cases.[5]

It is undeniable that cultural transmission depends on comprehension at almost every juncture. We human beings are just not in the habit of copying formulas we don't understand and then passing them mindlessly on to our neighbors. This in itself is not a fatal blow to the proposed Darwinian theory of cultural evolution because the intelligent agents active at these junctures are not miraculous. They are themselves products of earlier mindless evolution; cranes, not skyhooks. Moreover, the comprehension they exhibit, even in extreme cases, is typically insufficient to account for the cultural patterns their many attempts at copying and transmission eventually yield.

Just as genetic engineers, for all their foresight and insight into the innards of things, are still at the mercy of natural selection when it comes to the fate of their creations (that is why, after all, we are so cautious about letting them release their brainchildren on the outside world), so too the memetic engineer, no matter how sophisticated, still has to contend with the daunting task of winning the replication tournaments in the memosphere. One of the most sophisticated musical memetic engineers of the age, Leonard Bernstein, wryly noted this in a wonderful piece he published in 1955 in the *New York Times* entitled: "Why don't you run upstairs and write a nice Gershwin tune?" Bernstein had credentials and academic honors aplenty in 1955, but no songs on the Hit Parade.

A few weeks ago a serious composer-friend and I . . . got boiling mad about it. Why shouldn't we be able to come up with a hit, we said, if the standard is as low as it seems to be? We decided that all we had to do was to put ourselves into the mental state of an idiot and write a ridiculous hillbilly tune. (Bernstein 1959: 52)

They failed—and not for lack of trying. As Bernstein wistfully remarked, "It's just that it would be nice to hear someone accidentally whistle something of mine, somewhere, just once" (1959: 54).

His wish came true, of course, a few years later in 1961, when *West Side Story* burst into the memosphere. Leonard Bernstein was a brilliant, comprehending, foresighted, ambitious creator of musical designs that he hoped would replicate like viruses in brains around the world. He succeeded in a few cases, but so did many musically ignorant, lackadaisical, inadvertent exuders of equally infectious melodies. Other unforgettable melodies have no identifiable composer at all, but have emerged from untold rounds of differential replication. A theory that can encompass, and ultimately explain, all such varieties of cultural production will need to track the differential ability of authorless and authored items to get people to harbor them and pass them on, with or without comprehension.

Notes

1. The demands of this minimal Darwinism are far from trivial, and the ferocity with which Darwinian accounts of the evolution of language and sociality are attacked by some critics from the humanities and social sciences shows that mere consistency with evolutionary theory is not yet an accepted constraint in many quarters. This is a fact of life that we must deal with; fear of the thin edge of the wedge misleads many who hate the idea of a strong Darwinian theory of cultural evolution to resist conceding even consistency with evolutionary theory as the obvious requirement it is.

2. Such organisms need not be deemed to be making conscious decisions, of course, but the rationality, such as it is, of the "decisions" they make is typically anchored to the expected benefit to the individual organism. See, e.g., McFarland's (1989) distinction between an organism's goal function and its cost function, and Dennett (1989).

3. One might be tempted to treat the tolerance for variant "spellings" for proteins—e.g., there are over a hundred different ways of "spelling" lysozyme—as a sort of higher-level norm correction, but this is not strictly parallel since the copying at each locus is by local spelling, without ad lib interchange, except for mutations.

4. Following Sperber and Wilson's (1986) reasoning in a different domain 1986 we can note that no complicated ("Gricean") reasoning is required by the individual vectors in the series of transmissions, since they need not reconstruct the hidden recipe behind the production but simply use the optimality assumptions built into the intentional stance to home in on the intended production. It may often be difficult to reverse engineer the recipes for cultural products (styles of pottery, for instance) (Boyd and Richerson, 2000), but it is not typically necessary, since the intented properties of the result can be read off so readily.

5. There is considerable debate among memeticists about whether memes should be defined as brain structures or as behaviors, or as some other presumably well-anchored *concreta*, but I think the case is still overwhelming for defining memes abstractly, in terms of information worth copying (however embodied) since it is the information that determines how much design work or R and D does not have to be re-done. That is why a wagon with spoked wheels carries the idea of a wagon with spoked wheels as well as any mind or brain could carry it.

References

Bernstein, L. 1959. Why don't you run upstairs and write a nice Gershwin tune? Reprinted in *the joy of music* (pp. 52–62).

Borges, J. L. 1962. Pierre Menard, author of the *Quixote*: Labyrinths: Selected Stories and Other Writings. New York: New Directions, pp. 75–88.

Boyd, R., and Richerson, P. J. 1985. *Culture and the evolutionary process*. Chicago: University of Chicago Press.

Boyd, R., and Richerson, P. J. 2000. Memes: Universal acid, or a better mouse trap? In R. Aunger (ed.), *Darwinizing culture* (pp. 143–162). Oxford: Oxford University Press.

Dawkins, R. 1999. Foreword to Susan Blackmore, *The meme machine*. Oxford: Oxford University Press.

Dennett, D. C. 1971. Intentional systems. In D. Dennett (ed.), *Brainstorms. Philosophical essays on mind and psychology* (pp. 220–242). Cambridge, Mass.: MIT Press.

Dennett, D. C. 1987. *The intentional stance*. Cambridge, Mass.: MIT Press.

Dennett, D. C. 1989. "Cognitive ethology: Hunting for bargains or a wild goose chase?" In A. Montefiore and D. Noble, (eds.), *Goals, no-goals and own goals: A debate on goal-directed and intentional behaviour* (pp. 101–116). London: Unwin Hyman.

Dennett, D. C. 1995. *Darwin's dangerous idea: Evolution and the meanings of life*. New York: Simon & Schuster.

Feldman, M. W., and Cavalli-Sforza, L. L. 1981. *Cultural transmission and evolution: A quantitative approach*. Monographs in population biology 16. Princeton, N.J.: Princeton University Press.

McFarland, D., 1989. The teleological imperative. In A. Montefiore and D. Noble (eds.), *Goals, no-goals and own goals: A debate on goal-directed and intentional behaviour* (pp. 211–228). London: Unwin Hyman.

Powers, R. 1992. *The gold bug variations*. New York: Harper Perennial.

Sperber, D. 2000. An objection to the memetic approach to culture. In R. Aunger (ed.), *Darwinizing culture* (pp.163–173). Oxford: Oxford University Press.

Sperber, D., and Wilson, D. 1986. *Relevance: A theory of communication*. Cambridge, Mass.: MIT Press.

Williams, G. C. 1966. *Adaptation and natural selection*. Princeton, N.J.: Princeton University Press.

7   Conceptual Tools for a Naturalistic Approach to Cultural Evolution

Dan Sperber

While most anthropologists are not naturalistic in their approach and are not even trying to be, the project of a natural science of society and culture has always haunted the field. Radcliffe-Brown for instance wrote: "I conceive of social anthropology as the theoretical natural science of human society, that is, the investigation of social phenomena by methods essentially similar to those used in the physical and biological sciences" (1952: 189). It is not enough, of course, to want one's science to be naturalistic for it to be so. Radcliffe-Brown's anthropology remained well within the traditional social sciences of its time, quite removed from the natural sciences. Part of the reason is that what makes a science natural is not just its methods but also its ontology, i.e., the kinds of phenomena it recognizes as being part of the world and tries to account for. For instance, modern economics uses mathematical methods that are clearly scientific, but the objects it recognizes—markets, money, rational agents, etc—do not in any clear way belong to the furniture of the natural world. Their causal powers are not related in any clear or even hazy way to their material property and therefore economics is not (and is not intended as) a natural science.

Recently, several eminent biologists have approached cultural phenomena from an evolutionary perspective (Cavalli-Sforza and Feldman 1981; Dawkins 1976; Lumsden and Wilson 1981). Their contributions have stirred intense interest and discussion. Their goal and their methods have, of course, been naturalistic—much more clearly so than Radcliffe-Brown's—but their ontology is largely borrowed from the existing social sciences and is not clearly naturalistic. Religion, ritual, law, kinship, marriage, taboo, prestige, power, and so on, are the kind of things students of culture try to explain, whether they come from the traditional social sciences or from the biological sciences. There is no available naturalistic characterization of these things, nor is there any obvious way to provide one.

What do you have to do, then, to become really naturalistic? The presupposition of a naturalistic approach is that whatever has causal powers has them by virtue of its material properties. The first rule to follow is: Don't recognize phenomena unless your grasp of their material mode of existence justifies your attributing them with causal powers. The second rule is: Don't make a causal claim unless you can back it with the description of a mechanism, a description fine-grained enough for it to be reasonable to ask neighboring natural sciences to fill in the missing parts. These rules are well respected in fields such as biology, ecology, or geomorphology. In the study of culture, they are not understood, let alone respected, not even by people whose background and goals are unquestionably naturalistic.

To develop a naturalistic approach to culture, we must reconceptualize the field. To do so, we may draw inspiration from a science that is at once social and natural: medical epidemiology. In epidemiology, social macrophenomena such as endemic and epidemic diseases are analyzed in terms of patterns of microphenomena of individual pathology and interindividual transmission. In other writings, I have developed the project of a cultural epidemiology, and I have compared it with other approaches (Sperber 1996, 2001a). Here, I would like to characterize some of the most basic conceptual tools needed to develop a naturalistic ontology of cultural things in this epidemiological perspective. To help suggest how this conceptual rethinking may also be relevant to traditional anthropological pursuits, I will use an ethnographic example derived from my fieldwork among the Dorzé of South Ethiopia.

Among the many ways of explaining and coping with misfortune, two types deserve special attention, both for their worldwide recurrence and for their sociocultural import: mystical aggression or witchcraft, and mystical sanctions resulting from the transgression of taboos. In both types, misfortune is seen as initially caused by a human agent. In the case of mystical aggression, the culprit and the victim are distinct and indeed hostile individuals (or groups). In the case of transgression, the culprit and the victim are one and the same individual (or group). Many societies, while acknowledging both mystical aggression and transgression as possible explanations of misfortune, greatly favor one type of explanation over the other. This difference in the ascription of responsibility is rich in moral and social implications. For instance, in a witchcraft-oriented society, personal enrichment is likely to be viewed as evidence of guilt and therefore to be discouraged, whereas in a taboo-oriented society, it is likely to be viewed as a evidence of moral worth and to be encouraged.

The Zandé are a paradigmatic case of a society in which mystical aggression is the preferred explanation of misfortune (Evans-Pritchard 1937). The Dorzé were, when I visited them some 30 years ago, extreme in their preference for explanations in terms of transgression. When a misfortune occurred, most Dorzé would ask as a matter of course: "From which *gomé* did the misfortune come?" The term *gomé* denotes both the act of transgression and the resulting mystical sanction.

Dorzé adults could list hundreds of rules, the transgression of which would be *gomé*. Here are a few examples: It is *gomé* to let a drop of human blood fall in the food, to cook over a fire where a lizard has died, to ride a dog, to kill a snake, to have intercourse with a tanner or a potter (except, of course, for tanners and potters), to sacrifice an animal when one's father is alive, or to commit perjury. The rules are so many that everyone is likely to have been wittingly or unwittingly guilty of several transgressions. This does not seem to worry people. The problem of establishing that a particular transgression took place arises only when diviners are consulted, either because of a misfortune or in order to ascertain whether a sacrifice has been successful.

Only diviners are expected to know all the different types of *gomé* and all the ritual practices that must be followed to expiate a transgression. These diviners are of two main types: There are enteromancers who can "read," from the entrails of a sacrificed sheep or goat, which transgressions have been committed by the sacrificer or his dependents. This is a form of knowledge acquired with experience and typically is held by senior men who were themselves sacrificers. After having performed a sacrifice, a man typically will show the entrails to one or several senior neighbors who are competent in enteromancy. The second category of diviners consists of seers who use a variety of techniques, the most common one being geomancy. More important than the techniques they use is the seers' special divinatory gift, often linked to possession by a spirit. Seers can be men or women, central or marginal members of the community. People will often visit a seer who is some distance from their home. Unlike consultations of enteromancers, consultations of seers tend to be private and discreet affairs.

The ensemble of representations and practices involving the notion of *gomé* could be described by some traditional anthropologists as a cultural system and a major component of the Dorzé worldview. It could be described by others as a system of norms that shapes social relations and helps maintain social cohesion and power structures. Both types of macrolevel description would be insightful, and I do not particularly want to argue against them. It should be obvious, however, that neither the cultural nor the functional-structural approach is naturalistic. Do the ethnographic data also lend themselves to a more naturalistic approach, and could they provide relevant evidence in a naturalistic science of society and culture? To try and answer this, we must look—or here just peek—at the data at a lower, much more concrete level.

Ideas involving *gomé* and related practices are deployed in interindividual interactions, and in particular in consultations of diviners and in ritual practices. In spite of their variety, these interactions tend to follow a general pattern that can be represented in diagrammatic form (figure 7.1). The various possible sequences in this diagram can be illustrated with three chains of events that took place in Albazo's household over a period of 5 months (names have been changed; for a more detailed discussion, see Sperber 1980). His was one of a sample of forty households, the ritual activities of which were followed for 7 months in 1970–1971 (in collaboration with Judith Olmstead, who was surveying these households in a more systematic manner, and with her assistant Abesha Alemu; see Olmstead 1974, 1975). Albazo was at the time a 35-year-old weaver. He had spent several years working in Addis Ababa and had

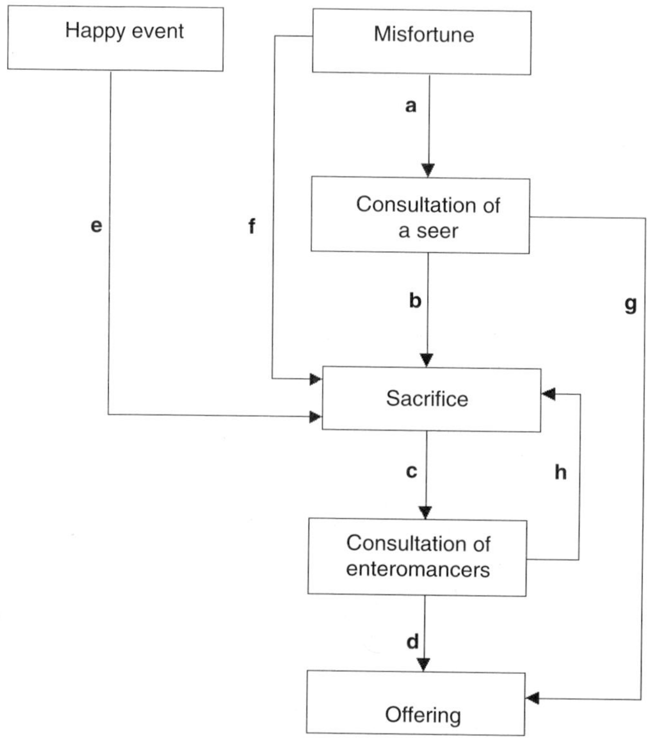

Figure 7.1
Possible chains of events in Dorzé divinatory and sacrificial practices. The most typical chain of events begins with a misfortune that causes the victim or the household head to consult a seer (a). The seer diagnoses a *gomé* and prescribes a sacrifice (b). The entrails of the sacrificed animal are shown to enteromancers (c), who prescribe an offering of beer or honey to put an end to the case (d). In other chains of events, the initial sacrifice may be caused by a happy event (e) or directly by a serious misfortune (f). For a minor *gomé*, a seer may directly prescribe an offering (g). When the entrails are "bad," enteromancers may prescribe a second sacrifice (h).

come back a year before at the death of his father. Present in his compound were his mother Bodé, his wife Maté, and a young sister's daughter who helped with domestic chores. Albazo's younger brother Abaté had remained in Addis Ababa to work. Albazo and Maté were without children; an infant son had died a few years earlier. Albazo was well off and could have felt quite contented if it had not been for his being child-less, and also for his mother not quite acknowledging that he was now the head of the household and persisting in treating him like a child.

Chain of events 1   In September 1970, at the time of the Ethiopian Maskal festival, Albazo sacrificed a lamb, saying: "Oh Maskal, you who have led me happily until now, be thanked!" He showed the entrails to three enteromancers of the neighborhood. They said: "There is a *gomé* of mother's insults." Albazo recognized that his mother Bodé had indeed insulted him because he had bought clothes for his wife Maté and not for her. The enteromancers instructed Albazo to make amends to his mother and to end the *gomé* with a libation, and so he did. This is a case of a sacrifice caused by the happy event of Maskal resulting in a consultation of diviners, the identification of a minor *gomé*, and an offering to end it (figure 7.1, path e-c-d).

Chain of events 2   In October 1970, Albazo's wife, Maté, whose eyes had been hurting her for several days, went to consult a seer. The seer, a geomancer, looked at the pebbles and said: "There is a *gomé* of honey." Maté remembered having eaten some of the honey her husband kept for offerings. Instructed by the diviner, she confessed her fault to her husband, who made an offering of honey. This is a case of misfortune resulting in the consultation of a seer and an offering. Neither sacrifice nor enteromancy is involved (figure 7.1, path a–g).

Chain of events 3   In January 1971, Albazo sacrificed a kid to his *k'ada ts'ala'e*, his "good luck demon," in order to find out why, unlike his friends, he was still childless. He showed the entrails to three enteromancers of his neighborhood. They said: "It is the *gomé* of the mother who gave you birth. She does not want you to have a child, and out of rancour, she cursed you. She should ask for forgiveness and give you a sheep [to sacrifice]." Bodé indeed confessed to being filled with bad feelings toward her son, and gave him a lamb. Albazo sacrificed the lamb and showed the entrails to the same three enteromancers. This time they said: "There is a *gomé* of you and your wife. Your *gomé* is to have said: "Don't let me have a child from her!" and her *gomé* is to have said: "Don't let me have a child from him!" Gather old men, have them forgive you both and make an offering of beer!" And so it was done. This is a case of a misfortune so serious that a sacrifice was performed in order to consult enteromancers, resulting in a second sacrifice and then an offering (figure 7.1, path f-c-h-c-d).

Before leaving aside, for the time being, the story of Albazo, let me explain what may have been its main import for the people involved. Albazo, his mother, and his wife were going through a transitory phase after the recent death of his father. Albazo's new position as household head, his descent, his age, his wealth, should all have concurred in progressively making him a well-established senior member of his community. However, he had no children and too much of a mother. The part played by the diviners must be understood against this background. They took advantage of a Maskal sacrifice, an eye complaint of Maté, and a sacrifice directly aimed at divination to ease the tension and redefine the roles in Albazo's household. Let the son be kinder to his mother, but let her acknowledge his authority; let the wife pay attention to her husband's new prerogatives; let the household head perform his new duties with serenity. Through divinatory procedures, the ineffectual anxieties caused by misfortunes are refocused on manageable psychological and social issues.

All this is, of course, anthropological data of a very familiar kind. In most cases, however (with notable exceptions such as Fredrik Barth 1975), such microlevel data are used to illustrate an explanation given in terms of macrolevel notions. What I want to suggest is that this microlevel is the proper level for naturalistic explanation. Now back to concepts.

Cognitive Causal Chains

Half a century ago, at the time when Radcliffe-Brown was calling for "a natural science of society," naturalistic explanations were not a genuine option in anthropology, nor more generally in the social sciences. The main reason for this has to do with the role that representations play in identifying the objects of the social sciences. It is indeed quite impossible to identify most, if not all, social-cultural phenomena without crucially relying on the mental representations of social agents. There is no theoretical perspective from which the *gomé* system, for instance, could be described without attending to general ideas the Dorzé have about *gomé* and the ideas individuals have about the specific cases in which they are involved in one capacity or another.

Until recently, there was no hint of a way to naturalize representations. More specifically, representations have material and abstract properties. Materially, public representations such as utterances or symbolic gestures may consist of marks on paper, or bodily movements, or any other kind of object in the environment that humans can produce and perceive. The material character of public representations is relatively unproblematic and poses no serious challenge to a naturalistic approach. Mental representations such as memories or desires consist of neural patterns in the brain. With recent developments in neurology, the material character of mental representations is beginning to be investigated in scientific terms. The most serious difficulty facing any attempt to naturalize representations has to do with their abstract properties. Representations, whether mental or public, have content, which is an abstract property. Moreover, it is by their content rather than their material properties that we tend to identify representations. For instance, we can talk of the tale of Goldilocks and the three bears without referring to its various material realizations, in speech, in writing, or in brain activation patterns. On the other hand, we would hardly ever find it of interest to talk of these public or mental material realizations without identifying them first as bearers of the content of Goldilocks and the three bears.

How can the abstract property of content be realized or implemented in the material world? How can the fact that abstract properties carry no causal power be reconciled with the fact that the content of a representation can be highly relevant in explaining its causal relationships? One thing that has greatly helped in answering these questions has been understanding how a computer program, which also has abstract content properties, can be materially realized and play a causal role in the world. With the recent development of the cognitive sciences—which is sometimes called the cognitive revolution—the goal of naturalizing representations is for the first time being approached in a realistic manner. We begin to understand how material processes systematically implement content relationships and have effects that are illuminated by these relationships. Let me sketch a brief and trivial illustration:

On October 31, at 7:30 p.m., Mrs. Jones's doorbell rings. Mrs. Jones hears the doorbell and assumes that there is somebody at the door. She remembers it is Halloween. She enjoyed receiving treats as a child and now as an adult she enjoys giving them. She guesses that there must be children at the door ready to trick-or-treat, and that if she opens it, she will be able to give them the candies she has bought for the occasion. Mrs. Jones decides to open the door and does so.

We have here an environmental change (the ringing of the doorbell), a process of perception (Mrs. Jones hearing and recognizing the doorbell), a process of epistemic inference (her inferring that there is somebody at the door), the retrieval from memory of a belief (that it is Halloween) and a desire (to give candies to children), a second process of epistemic inference (inferring that there must be children at the door ready to trick-or-treat), a process of practical inference (inferring that in order to fulfill her desire to give candies, Mrs. Jones should open the door) and the realization of an intention (to open the door), resulting in an environmental change (the opening of the door).

These events are causally related in a complex chain. This is a special kind of causal chain that I will call a cognitive causal chain, or CCC for short. What makes it cognitive is, roughly, the fact that to each of the causal links in the chain, there corresponds a semantic or content relationship. Mrs. Jones's perception of the doorbell ringing both represents the doorbell ringing and is in part caused by it. Mrs. Jones's remembering that it is Halloween and what is likely to happen now is similar in content (with appropriate updating) to the knowledge derived from previous experiences of Halloween, and that stored knowledge is among the causes of her remembering. Mrs. Jones's coming to specific conclusions (whether epistemic—*someone is at the door, children are at the door ready to trick-or-treat*—or practical—*let me open the door*) is both justified by specific premises and caused in part by her entertaining these premises. Mrs. Jones's opening of the door both satisfies her intention to do so and is caused in part[1] by this intention.

Semantic relationships such as truth, satisfaction, justification, or similarity of content are abstract relationships and not causal ones. Perception, inference, remembering, and the carrying out of an intention are causal processes. These processes, however, are characterized in terms of the abstract semantic relationships they tend to instantiate. When we describe mental processes as processes of perception, inference, remembering, or intending, we mean that these processes tend to produce outputs that are in a characteristic semantic relationship to their inputs. A successful perception yields a representation that represents the very stimulus that caused the perception. A successful process of inference yields a conclusion justified by its input premises. A successful remembering yields a memory similar to the initially stored information. The successful carrying out of an intention brings about the state of affairs represented in the intention.

Mental life is made up of CCCs where the links are both semantic and causal, and not fortuitously so, but because each of the causal processes involved has the function of instantiating a certain type of semantic relationships.[2] Materialists of the past could well postulate that the causal aspects of cognition should in principle be wholly describable in material terms, but it is only recently that we have become capable of actually describing material mechanisms that instantiate abstract semantic relationships. When describing CCCs, not only can we claim, on general grounds, that they occur in the brain and in the interactions between the brain and its environment, we can also begin to describe in computational and neurological terms the kinds of material processes that realize these CCCs.

Assuming that the cognitive sciences do provide us with a naturalistic notion of mental representations (or at least with a notion that is in the process of being naturalized), how does this help us naturalize the notion—or notions—of representations used in the social sciences? Psychologists talk about individual mental representations. Social scientists talk about representations that are in some sense collective (whether they use the term *representation* or just talk of ideologies, beliefs, values, and so on, which are all kinds of representations). It could be argued then that "representation" in psychology and "representation" in the social sciences merely share the most basic property of representations in general, i.e., "aboutness," being about something, having some "content," but that otherwise they are quite different things.

Social Cognitive Causal Chains

The story of Mrs. Jones, as told so far, is typical of individual psychology. It is all about inputs to an individual organism, its internal processes, its individual representations, and the behavioral outputs of this organism. In this particular case, however, the causal chain directly involved other individuals, to begin with, Billy and his little sister Julia:

Billy and Julia are following the Halloween practice of going from door to door in the street, hoping to be given candies. When they reach Mrs. Jones's door, Billy rings the bell with the intention of letting the house owner know that someone is at the door, and of making her open the door . . . [insert Mrs. Jones's story as told earlier]. Mrs. Jones opens the door. Billy and Julia shout "trick or treats!" Mrs. Jones gives them candies.

Ringing a doorbell is a process of communication. Like all processes of communication, it has the function of causing, in the mind of the addressee, the formation of a representation similar in content to the representation the communicator had in mind (in this case, the content is that the addressee should open the door to the ringer of the bell). Notice that in such an interindividual causal chain, the interindividual links are no less cognitive (i.e., instantiating semantic relationships) than the intraindividual ones. Communication instantiates semantic relationships of similarity of content, not within an individual, but across individuals. When a CCC extends over several individuals, I call it a social CCC. Social CCCs may involve just two individuals, or a few, or extend indefinitely over social time and space. Thus the interaction between Mrs. Jones and the children on the night of Halloween is just a fragment of a much longer and wider social CCC that links all particular Halloween events to the emergence of the practice and to one another.

Communication provides paradigmatic examples of social CCCs. In the case of an assertive act of communication, the social CCC typically goes from a mental event in the communicator, to an environmental event (e.g., the production of a signal such as a doorbell ring, or of a linguistic utterance), to a mental event in the addressee, and stops there. In the case of a request, the social CCC typically extends one step further, to a second environmental event that fulfills the request. Thus Mrs. Jones, having understood that someone wants the door opened, opens the door. Both Billy and Mrs. Jones form the intention that the door be opened. However, while Mrs. Jones is in a position to carry out this intention by herself, and does so, Billy, for the same purpose, needs to recruit Mrs. Jones and does so by communicating a request. Mrs. Jones's fulfillment of Billy's request instantiates a semantic relationship between one individual's mental state and another individual's action. More generally, the fulfillment of a desire by means of a request to another individual is a major kind of social CCC. This is true of the very simple communication established by ringing a doorbell as well as the more elaborate back-and-forth communication involved in ongoing collective action.

While communication provides the most obvious cases, noncommunicative forms of interaction may also determine social CCCs. These include imitation and other forms of emulation. Consider a group of people walking for the first time from a new settlement to some landmark in the distance. One person walks in front, choosing the best path through the bush—a cognitive process—and treading over grass and ground. The others follow in line, each contributing to marking the path. The following days, months, and years, when people follow this footpath they will each contribute to maintaining it as a stable and salient feature of the landscape, causing others, or themselves on later occasions, to borrow it in turn. The path started its existence as the visible effect of a series of microdecisions (of stepping here rather than there) of one individual. This visible effect caused other individuals to make similar microdecisions, adding to the initial effect.

Now the path has become the collective production of all those who have followed it, an item of the socially shared landscape, and a spatially extended perceptual input guiding the steps of every new walker. There is, then, a social CCC going from the microdecisions of past walkers to those of future walkers, via the environmental changes that each contributes. At times, as when individuals walk in line, there may be a deliberate imitation of the behavior of one individual by others. A solitary walker, on the other hand, may choose to follow a path without paying attention to the fact that in doing so she or he is emulating other people. Whether conscious or unconscious, such spontaneous forms of emulation may determine a social CCC, and do this without resorting to communication proper.

Mental Representations and Public Productions (Including Public Representations)

Social CCCs link mental and public things. The mental things involved are mental representations and processes, which may cause behaviors that alter the environment in ways that can be perceived and that can serve as stimuli to further cognitive processes. Some of these environmental changes are perceptible as processes, e.g., bodily movements, speech sounds; others are perceptible as stable states of the environment, e.g., the presence of paths, buildings, artifacts, or writings. I will call all such perceptible behaviors and effects of behavior public productions. Some public productions, for instance utterances, signals, or pictures, are produced for the purpose of being perceived and causing mental representations. These public representations form a particularly important subclass of public productions. Social CCCs, then, are characterized by an alternation, along the causal chain, of mental representations and public productions (including public representations).

The three chains of events in Albazo's story were each a case of a social CCC. The point of saying this is not to introduce new terminology for terminology's sake. It is to suggest a level at which the very different ingredients of such a causal chain—worries, misfortune, divination, confession, sacrifices, offerings, and so forth—can be seen as an alternation, along the causal chain, of public productions and mental representations that are linked by causal relations and by content relations. The mental representations involved were beliefs and desires both caused and justified by public events, and most of the public productions were, in this case, public representations such as utterances and symbolic gestures fulfilling mental intentions and caused by these intentions.

In the diagram I presented earlier (figure 7.1), outlining the various kinds of *gomé*-related chains of events, only public events were mentioned. However, public events cause further public events through the mediation of mental events, which must also be taken into consideration. To illustrate this, let us go back to the second chain of events. Albazo's wife, Maté, suffering from eye pain, consults a seer. What are the psychological processes that link her eye pain to her going to consult a seer? Not all such pains result in such a course of action. Maté might have sought help in traditional medicine or she might have waited for the pain to go away. However, her husband's mother, Bodé, a few days earlier had scored a kind of domestic victory. After Albazo had sacrificed a thanksgiving lamb, the enteromancers had diagnosed a *gomé* caused by his having bought clothes for his wife but not for his mother. By going to the seer, Maté makes sure that her husband's next ritual action will be for her sake. In other words, Maté's capacity to anticipate some indirect effects of her action may well be playing a decisive role.

When the seer diagnoses a *"gomé* of honey,"* Maté could interpret this in various ways. She could for instance have wondered whether she had unwittingly spoilt honey or mead. Or she could reject the seer's diagnosis by saying that she did not see what he could be referring to. In this case, however, the words of the seer cause her to remember having eaten from her husband's ritual honey. By interpreting the seer's diagnosis as referring to such an event, Maté turns the fault to which she will have to confess into a reassertion of her husband's privileges as the new head of the household. After the consultation, Maté could have decided to ignore it altogether or perhaps to see another seer. Similarly, her husband Albazo could have dismissed the whole matter. Each of these microdecisions would have changed the chain of events. Thus such a chain of events cannot be explained just by saying that it conforms to a cultural pattern or norm. On the contrary, the cultural pattern is a recurrent one—is a pattern—because relatively idiosyncratic causal factors tend, in a variety of circumstances, to converge on similar courses of action.

More generally, at every juncture in every social CCC, the mental processes of the individuals involved may tilt the chain of events one way or another. These mental processes exhibit cross-individual regularities. Some of these regularities have to do with basic cognitive and emotional dispositions that are part of the biologically evolved psychological makeup of humans. Others are contingent on historical and local circumstances. The anthropologist's goal is not to explain individual cases, but recurring patterns. However, I have argued that explaining recurring patterns requires attending to the kinds of psychological factors that affect individual cases.

Cultural Cognitive Causal Chains

Most social CCCs are short. They bring about only local and brief transfers of information, coordinations of behaviors, or movements of matter such as transfers of goods. They are episodes like the three chains of events in Albazo's household. Although they are causally related to one another, each has its own specific content. Some social CCCs, though, are long and lasting, involve a great many individuals over

Table 7.1
Types of causal chains

Cognitive causal chain (CCC)
A causal chain in which each link instantiates a semantic relationship

Social cognitive causal chain (Social CCC)
A CCC that extends over several individuals

Cultural cognitive causal chain (CCCC)
A social CCC that stabilizes mental representations and public productions in a population
and its environment

---

time, and exhibit no discontinuity of content. The Halloween interaction I evoked was a typical fragment of such an extended chain. These long and lasting social CCCs have the effect of stabilizing mental representations and public productions in a population and its environment. Mental representations and public productions (practices or artifacts) that are stabilized by such extended social CCCs correspond to what we call the cultural. I propose to call social CCCs that stabilize cultural representations and productions cultural cognitive causal chains, or CCCCs for short (table 7.1).

Let me illustrate what I mean when I say that representations or practices are "stabilized." Take the case of a folktale such as Goldilocks and the three bears, and take it at the time when it was only transmitted orally. Each time the tale was told it contributed to the audience's knowledge of the tale and to their desire to hear it again and possibly to tell it in turn. If it had not done so to a sufficient degree, the tale would not have remained as a stable cultural representation, since it was stabilized only by the CCCC that linked tellings of the tale (public productions, and more specifically, public representations) to individuals' knowledge of the tale and motivation to tell it in turn (i.e., mental representations).

The existence of CCCCs and their stabilizing effect are among the most obvious aspects of human social life, but they are not easily explained. Human memory, imitation, and communication are not true replicating mechanisms. Their outputs are rarely, if ever, identical to their inputs. Even when the alterations between, say, the story heard and the story understood, the story understood and the story remembered, the story remembered and the story told are small—and often they are large—the cumulative effect of these alterations in an extended social CCC is likely to be such that contents rapidly decay or are transformed beyond recognition. This is indeed what happens with most stories told. For instance,

Carol tells Bob how she made a fuss at the supermarket. Bob tells Ted how Carol made a fool of herself at the supermarket. Ted, a while later, mixes this story with another one he had heard about Carol at the library, embellishes it and tells it to Alice, who does not believe it anyhow, and ends up remembering only that Ted accused women of behaving absurdly in department stores.

Most social CCCs are like these interactions among Bob, Carol, Ted, and Alice; they don't extend very far and they stabilize very little if anything at all.

Only some mental representations such as folktales, and some public productions such as sacrificial rites, exhibit great resilience and do get stabilized by CCCCs. That is, they remain recognizably similar to antecedent representations or productions in the chain. Recognizable similarity is a matter of degree. There is no real boundary, therefore, between unquestionably cultural representations such as Goldilocks on the one hand, and apparently idiosyncratic stories such as that told by Carol to Bob about her adventures at the supermarket. Even the latter is recognizably similar, in its gist, to stories often told. In telling it, Carol was relying not just on her memory of the event but also on her memory of similar stories she had heard. In retelling it, Bob, and then Ted, were altering it, not at random, but in the direction of the cultural cliché that Alice all too easily recognized. No social CCC is ever unconnected to cultural CCCs; rather, all short and local social CCCs are offshoots of one or several cultural CCCs, and these offshoots may contribute to the persistence of the CCCCs themselves.

To further illustrate this point let us go back to Albazo's story. The three chains of events I described were clearly idiosyncratic versions of enduring patterns, offshoots of CCCCs that criss-cross Dorzé social life. When Albazo decided to sacrifice a thanksgiving lamb for the festival of Maskal, for instance, this decision and his action were clearly linked, both causally and in content, to innumerable similar decisions and courses of action taken in the past by other Dorzé household heads (and in particular by Albazo's father). Again, in showing the entrails to enteromancers of the neighborhood, Albazo was reproducing countless past actions by Dorzé household heads. On the other hand, in apologizing to his mother for having bought clothes for his wife but not for her, Albazo was attending to the particulars of his situation; nevertheless, his behavior was recognizably similar to that of many others.

The diviners themselves, when "reading" the entrails, were producing a version of past diagnoses adjusted to the particulars of the situation. Their thought processes and their diagnosis were at the crossing point of two social CCCs: the short social CCCs triggered by Albazo's Maskal sacrifice and the long cultural CCC that stabilizes the particular type of *gomé* that they diagnosed.

One point should be underscored here that is highly relevant to the explanation of cultural resilience and change. The diviners were extremely unlikely to opt out, so to speak, of available CCCCs, and to produce a truly novel diagnosis—a new type of transgression for instance—that could have been challenged by Albazo or by other ritual experts. Still, there was a wide range of types of diagnoses to choose from. Each particular type is maintained by a specific cultural chain. In choosing a particular diagnosis, the diviners are contributing to the persistence of one of these cultural chains. Each time a type of diagnosis is chosen, it gains in salience and in the likelihood of being considered on future occasions. If a particular type of diagnosis becomes more and more popular with the enteromancers, its cultural importance will grow, and so will the likelihood that this subvariety of diagnosis will become distinguished, leading to a split of the underlying CCCC into several new CCCCs. On the other hand, if a type of diagnosis becomes reproduced less and less, its CCCC will lose momentum and may eventually come to an end.

The evolution of the *gomé* system is thus to a large extent determined by the mental processes and the interactions that on each particular occasion, tilt the diviners' diagnosis one way or another. Among the factors that contribute to the diviners' preferences, I would like to mention two: the reactions of the persons consultanting them and the state of the entrails. The persons seeking help are more or less willing to accept different diagnoses. They may, like Albazo with the enteromancers or Maté with the seer, recognize without difficulty that they are guilty of a transgression of the type mentioned by the diviners. In elaborating on the diviners' diagnosis, they contribute to the way the diviners themselves understand and mentally exemplify their somewhat cryptic diagnoses, such as "*gomé* of mother's insult," or "*gomé* of honey." The persons receiving a diagnosis may also be sceptical, or even disbelieve it. Diviners who produce unconvincing diagnoses may readjust their interpretations, or else they are likely to be less consulted in the future, and therefore play a less important role in cultural transmission.

Diviners practicing enteromancy are also constrained, in their diagnosis, by the state of the entrails they are asked to read. After all, there are rules of interpretation, and different shapes, spots, and anomalies of the entrails have more or less standard interpretations. Different oddities of entrails have different frequencies, over which the sacrificers and the enteromancers have no control. However, rules of interpretation are themselves cultural representations maintained by their own CCCCs. It is likely that without their awareness, the interpretive preferences of the enteromancers determine the evolution of the rules of interpretation. If, say, at a certain historical time, the swelling of a certain gland in the entrails is taken to indicate a type of *gomé* that diviners are less and less inclined to diagnose, and if this swelling is relatively frequent, it is likely that through a series of microdecisions, the interpretation of this swelling will be altered. The frequent swelling will progressively be interpreted as indicating a favored type of *gomé*.

I introduced the Dorzé example by contrasting those societies that generally explain misfortune in terms of witchcraft, and those that, like the Dorzé at the time of my visit, almost exclusively resort to explanations in terms of transgression and sanction. The relative place given to these two types of explanations results from a series of microdecisions and behaviors along cultural causal chains. The Dorzé did recognize various forms of mystical aggression—*bitha* or "sorcery" in particular—as possible sources of misfortune. It is just that these were rarely invoked. They were never diagnosed by enteromancers and only rarely by seers. After the 1974 Ethiopian revolution when the Emperor Haile Selassie was deposed and replaced by a Marxist leadership, the senior members of Dorzé communities—including most of the enteromancers—were often denounced as bourgeois. Many rules, the transgression of which was considered *gomé*, were denounced as reactionary. In particular this was the case of rules having to do with seniority and ritual prerogatives. These are types of transgression typically "read" from the entrails, where patterns of blood vessels are interpreted as a genealogical tree indicating relationships of seniority and their possible disruptions. Such changes in turn rendered enteromancy less attractive than other forms of divination; they made people less willing to accept diagnoses involving issues of seniority. They encouraged seers to prefer making diagnoses in terms of other kinds of *gomé* (having to do, for instance, with food, sex, or possession by a spirit), and to increase the frequency of diagnosis in terms of *bitha*, sorcery. Although the new ideology was equally against ideas of *gomé* and *bitha*, both of which were denounced as superstitions, its propagation had in fact the effect of favoring explanations of misfortune in terms of mystical aggression rather than in terms of the transgression of taboos.

More generally, the existence of all degrees of balance across human societies between explanation of misfortune in terms of witchcraft or in terms of taboo is the cumulative effect of microprocesses, both mind-internal and mind-external, along the causal chains of culture. While a task for ethnography is to describe the factors that locally stabilize or alter the balance one way or the other, a task for a naturalistic anthropology is to identify the types of factors that may be involved in such stabilization or changes, and to explain how these factors work by affecting people's minds and environments.

Conclusion

Social CCCs are not an aspect of the social. They *are* the social. Things are social to the extent that they are involved in cross-individual cognitive causal chains. Cultural causal chains are not an aspect of the cultural. They *are* the cultural. Social things are cultural to the extent that they are involved in cultural cognitive causal chains. I know of no counterexamples to these claims. On the contrary, I believe they provide a fine-grained way to tease apart what is social and what is not, and within the social, what is cultural and what is not. Moreover, thinking of cultural and social things in those terms—as causal articulations of mental and environmental events and states—should permit the development of a naturalistic ontology of the social.

Social scientists might be concerned less with problems of ontology and conceptual analysis, and more with substantive matters, and in particular with the place given to mental things in an epidemiology of representations. They may feel that I give far too much importance to representations and cognition in characterizing the social and the cultural; or worse still, that I am reducing the social and the cultural to the mental. Aren't agriculture or war, for instance, paradigmatic examples of things social? Aren't artifacts and public performances paradigmatic examples of things cultural? Yet, without denying their cognitive dimension, surely these are not principally mental things, and their importance has to do primarily with their effects on the bodily lives—not just the mental representations—of people? I fully agree, and if this is thought to be an objection to the naturalistic approach I advocate, then I have failed to make myself understood.

Let me be quite clear. Many things can be caught in a web of social CCCs, not only mental and public representations, but also other public productions, such as paths, buildings, crops, markets, machines, and massacres. All things caught in a social CCC have mind-internal—or psychological—causes and effects, and all have mind-external—or environmental—causes and effects. Which of these causes and effects matter more may vary with specific cases and points of view. In the case of *gomé* chains of events such as those that took place in Albazo's household, much of the explanatory weight lies on the psychological side, even though typically these chains of events are triggered by nonpsychological events or states of affairs, such as a disease or a bad crop. In the case of a path for travel, the psychology is rather trivial and ecology plays a greater explanatory role. After all, in the absence of deliberate maintenance, the stability of paths in a community depends on some balance between the rate of plant growth and erosion on the one hand and the intensity of use on the other. The epidemiological approach must, in all cases, combine an environmental perspective and a psychological perspective and is not committed to—or opposed to—giving precedence to one or the other of these two perspectives.

Why then characterize social and cultural causal chains in term of their psychological links rather than their environmental links? To begin with I would like to stress that psychological links are themselves a subcategory of environmental links. They are links located in brains and bodies that are themselves part of the environment. So, to recognize a special place for psychological links in a social CCC is just to highlight one type of ecological factor. The reason for giving a defining role to psychological links is that the other, nonpsychological links in a social CCC can be indefinitely varied: sounds of speech, gunshots, images, paths, dances, foods, clothes, machines, and so on. No subcategory of these environmental links is either necessary or sufficient for the causal chain in which they occur to be thereby a social chain. What makes a causal chain social is the cognitive linking of different individual minds. What makes a social chain cultural is the (relative) stabilization of representations. It does not follow, however, that the psychological ingredients of the social are more interesting than its nonpsychological ingredients. Interest is a pragmatic matter.

Another concern some social scientists might have is that there is some arbitrariness in distinguishing the social and the cultural and recognizing each as an equally worthy object of study. They might argue that everything that is social is also cultural, and conversely. This is true in the human case, of course. However, in this respect, humans differ greatly from other social animals. Most social animals only transmit information about the here and now (e.g., "beware, there is a predator!"). Whatever knowledge and skills members of these animal populations durably share, they owe to similar biological dispositions expressed in the same environment, rather than to their ongoing mutual interactions. In other terms, the social CCCs of most social animals do not stabilize any common knowledge or skills; they are not CCCCs. Still, there are fascinating exceptions—examples of practices (and therefore of the mental representations that make them possible) spreading through imitation and other forms of social learning and stabilizing in nonhuman animal populations. For instance, it is now well documented that different chimpanzee populations have different, socially transmitted techniques—for termite fishing, for instance (McGrew 1992). These techniques are, in other terms, transmitted through CCCCs. They are cultural. Still, even in those nonhuman animals that exhibit some degree of cultural transmission, most activities, whether individual or social, are free of any cultural influence.

In the human case, and in the human case only, culture is all-encompassing. All social CCCs draw on culturally transmitted representations, even when they do not directly propagate them. The domain of the social and that of the cultural are indeed coextensive. In this extensional sense, there isn't any difference between social and cultural things. On the other hand, being social and being cultural are two different properties. Something is social to the extent that it involves some cognitively mediated coordination among individuals. Something is cultural to the extent that it involves the stabilization of representations or productions by means of cognitively mediated coordination among individuals.

One may be more interested in the social or in the cultural aspects of things that are inevitably both social and cultural. That is, one may be more interested in answering the question, "How do humans coordinate?" or the question, "How do representations and productions evolve and stabilize?" but the domain of facts relevant to answering these two questions is the same. I have tried to suggest that a naturalistic answer might be given to both questions. For this, the domain of the social sciences must be reconceptualized by recognizing only entities and processes of which we have a naturalistic understanding. These are mental representations and public productions, the processes that causally link them, the social and in particular the cultural CCCs that form these links, and the complex webs of such causal chains that criss-cross human populations over time and space. Start from such a reduced ontology, and, yes, the goal of a truly natural science of culture and society might not be entirely utopian.

Notes

An earlier version of this text was given as the Radcliffe-Brown Lecture in Social Anthropology 1999 and Sperber (2001b).

1. Since I am never describing the entire complex cause of some event, but only highlighting some part of it, from now on "caused" must be understood to mean "caused in part."

2. A number of philosophers—Fred Detske, Ruth Millikan, Karen Neander, and David Papineau for instance—have tried to naturalize meaning by appealing to a notion of function. Although no final and compelling solution has yet been found, I see these attempts as being obviously on the right track. For an overview, see Jacob (1997).

References

Barth, F. 1975. *Ritual and knowledge among the Baktaman of New Guinea*. New Haven, Conn: Yale University Press.

Cavalli-Sforza, L. L., and Feldman, M. W. 1981. *Cultural transmission and evolution: A quantitative approach*. Princeton, N. Y.: Princeton University Press.

Dawkins, R. 1976. *The selfish gene*. Oxford: Oxford University Press.

Evans-Pritchard, E. E. 1937. *Witchcraft, oracles and magic among the Azande*. Oxford: Clarendon Press.

Jacob, P. 1997. *What minds can do*. Cambridge: Cambridge University Press.

Lumsden, C., and Wilson, E. 1981. *Genes, minds, and culture*. Cambridge, Mass.: Harvard University Press.

McGrew, W. C. 1992. *Chimpanzee material culture: Implications for human evolution*. Cambridge: Cambridge University Press.

Olmstead, J. 1974. Female fertility, social structure, and the economy: A controlled comparison of two Southern Ethiopian communities. PhD dissertation, Columbia University, New York.

Olmstead, J. 1975. Land and social stratification in the Gamo highlands of southern Ethiopia. In H. Marcus (ed.), *Proceedings of the first U.S. conference on Ethiopian studies, 1973* (pp. 223–234). East Lansing, Mich.: Michigan State University Press.

Radcliffe-Brown, A. R. 1952. *Structure and function in primitive society*. London: Cohen and West.

Sperber, D. 1980. The management of misfortune among the Dorze. In R. Hess (ed.), *Proceedings of the fifth international conference on Ethiopian studies* (pp. 207–215). Chicago: Office of Publications Services, University of Illinois at Chicago Circle.

Sperber, D. 1996. *Explaining culture: A naturalistic approach*. Oxford: Blackwell.

Sperber, D. 2001a. An objection to the memetic approach to culture. In R. Aunger (ed.), *Darwinizing culture: The status of memetics as a science* (pp. 163–173). Oxford University Press.

Sperber, D. 2001b. Conceptual tools for a natural science of society and culture. *Proceedings of the British Academy* III: 297–317.

II Brain, Cognition, and Evolution

8 Brains, Cognition, and the Evolution of Culture

R. I. M. Dunbar

Although strong claims have been made for culture in animals (e.g., Whiten et al. 1999), it remains the case that only humans exhibit massive variation in behavior that can be attributed directly to culture (and the mechanisms of social learning that underpin its transmission). Thus even if chimpanzees and other primates exhibit considerable phenotypic variation at the behavioral level, the problem we have to explain is why humans should exhibit this capacity to such an extraordinary degree.

In this chapter I ask two questions. First, what makes it possible for humans (but not other apes or monkeys) to exploit cultural processes as extensively as they do? Second, I consider the more difficult question of why this capacity might have evolved.

Neural Substrates

It now seems fairly safe to conclude that, at least within primates, the evolution of brain size has been driven by social demands. Relative neocortex size correlates with social group size (Dunbar 1992, 1998; Barton and Dunbar 1997), suggesting that the demands of maintaining the cohesion of large highly bonded groups through time depends crucially on the computing capacity available to manage and manipulate the database on social relationships. This is borne out by the finding that other more subtle aspects of social behavior (including the size of grooming cliques, frequencies of social play, the ability to undermine the power-based strategies of dominant animals) also correlate with relative neocortex size (Kudo and Dunbar 2001; Lewis 2000; Pawlowski et al. 1998).

Humans seem to fit perfectly into the spectrum of nonhuman primates. If we use the nonhuman primate data to project a group size for modern humans based on our neocortex size, the figure we get (approximately 150 individuals) seems to correspond to a group size that is commonly found in both traditional (hunter-gatherer) societies and in modern industrial societies (Dunbar 1993). In hunter-gatherer societies, the value of 150 corresponds in size to the clan (e.g., as defined in Australian aboriginal society) or regional grouping (e.g., as defined in studies of !Kung San in southern

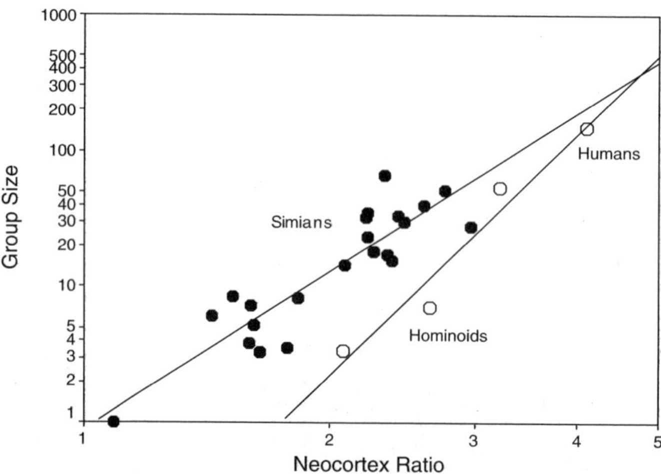

Figure 8.1

Mean group size plotted against relative neocortex size (neocortex volume divided by the volume of the rest of the brain) for anthropoid primates (monkeys and apes). Each data point is a representative species from one genus. Apes are shown separately from monkeys. Data are derived from the dataset collated by Stephan et al. (1981). (Redrawn from Dunbar 1992.)

Africa). In modern postindustrial societies, it seems to correspond to the number of individuals that you know well, the number of people you would not feel embarrassed about asking a favor of.

However, if we look more closely at the group size–brain size relationship in primates, an interesting point emerges. The great apes appear to lie on a plane or grade separate from that of the monkeys (figure 8.1). It looks suspiciously as though the apes need more computing power than monkeys do to sustain a social group of a given size. This might be taken to suggest that the ape lineage as a whole uses a different kind of social style than monkeys do. This possibility is given further support from an unexpected anatomical direction.

We know that different parts of the neocortex are more or less dedicated to different tasks (vision, hearing, sensation, motor control, etc.). If so, it seems intrinsically likely that the group size effect reflects not the influence of the whole neocortex but the activity of some more limited part. The logical place to look is perhaps the frontal cortex, since much of the neocortex posterior to this is fairly strongly dedicated to specific tasks.

Unfortunately, the primate brain database that we have available differentiates only the primary visual area (area V1) from the neocortex as a whole. Even so, a plot of social group size against each of these two parts of the neocortex suggests that

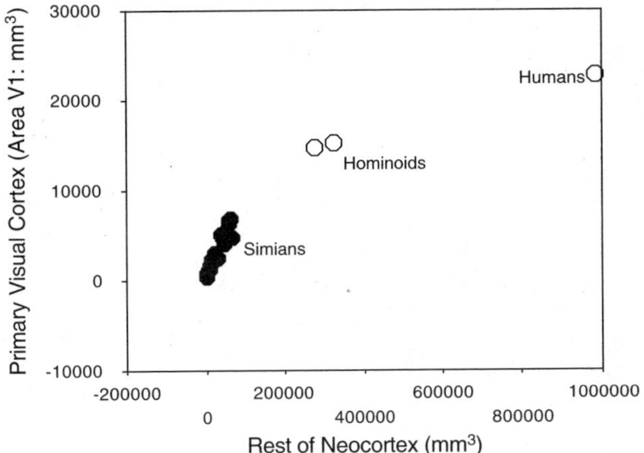

Figure 8.2
Volume of the primary visual cortex (Brodmann area V1) plotted against the rest of the neocortex for individual anthropoid species (each species representing a different genus). Humans are the data point on the extreme right. (Redrawn from Joffe and Dunbar 1997.)

the primary visual area plays only a limited role in determining social group size in primates (Joffe and Dunbar 1997). It really is something about the more frontal areas of the cortex that is important.

If we now plot the two parts of the cortex against each other, then something quite surprising emerges. Although there is a linear relationship between the visual area and the rest of the neocortex in smaller-bodied primates (principally monkeys), this relationship starts to level off in favor of larger nonvisual areas in the great apes (figure 8.2). This is difficult to understand unless we assume that the visual system is subject to diminishing returns in terms of visual acuity as processing area increases. In other words, adding extra cortical processing volume may not lead to a proportional increase in visual acuity after a certain point. The importance of this is that it provides spare computing capacity that can be devoted to other cognitive processes. This claim is supported by the fact that the optic tract is more or less linearly related to the area of the primary visual cortex (figure 8.3). Since visual processing efficiency is limited by the cross-section of the optic tract (in effect, the number of neurons that can link the retina with the visual processing areas in the brain), there is no point in having a primary visual area larger than is minimally adequate to process the incoming signals from the eyes.

The scale of this becomes more apparent if we use the near-linear relationship between the two brain components in simians as shown in figure 8.2 to predict the visual area for the apes, and then use these to calculate residual values from the actual

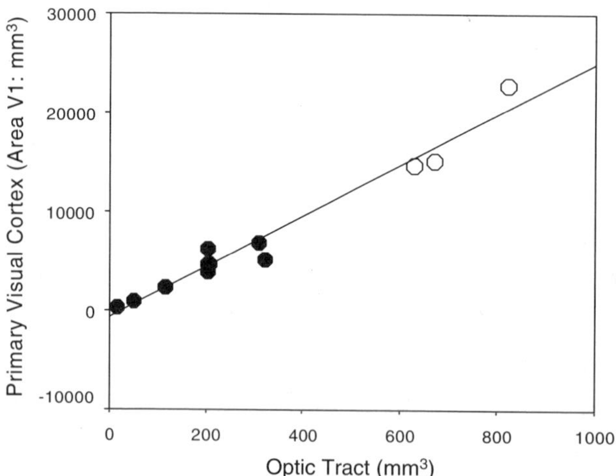

Figure 8.3
Volume of the optic tract plotted against the rest of the neocortex for individual anthropoid species (each species representing a different genus). Humans are the data point on the extreme right. (Source: Stephan et al. 1981.)

sizes of their primary visual areas (figure 8.4). The monkeys scatter either side of the line, much as would be expected. With the apes, however, we begin to see an exponential rise in the residual value. This is in effect the spare capacity freed from the requirements of (at least, primary) visual processing that can be used for other purposes. The importance of this will become clear in the following section.

Cognitive Mechanisms

It is not exactly clear how humans differ from other primates in terms of fundamental cognitive abilities. It is easy to see that humans can do things that other primates cannot (e.g., recite the Koran, deduce Euclid's geometric proofs), but it is not entirely clear why humans should be able to do this (other than the rather prosaic answer, of course, that their bigger brains allow them to). Exactly what does a bigger brain confer? One answer is the capacity to develop computationally expensive processes such as theory of mind (ToM).

Theory of mind is the ability to mind read or imagine how another individual sees the world. It is encapsulated in the statement: "I believe that you think the world is flat." In other words, it underpins the ability to imagine another individual's state of mind and in particular his or her belief state. Formal theory of mind in this sense is sometimes equated with second-order intentionality (on a scale where first-order intentional beings know their own mind states but no more). ToM appears to be crucial

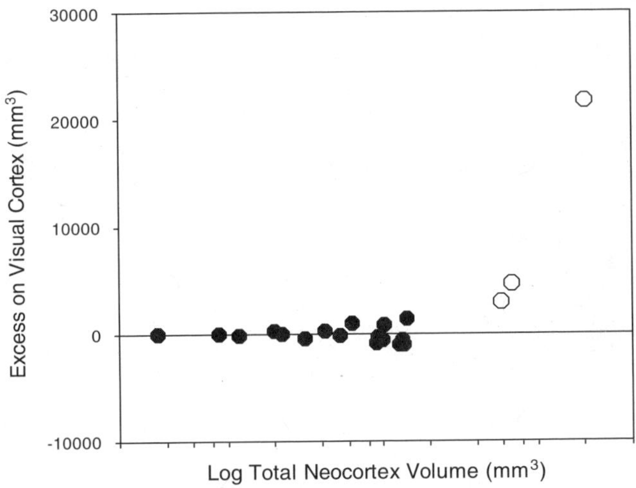

Figure 8.4
Relative volumes of spare neocortical capacity in different anthropoid primate species. The plotted values are the difference between the nonvisual (V1) neocortex predicted from the total neocortex volume by the relationship for simian primates (monkeys) and the observed value plotted against observed V1 volume for each species shown in figure 8.3. Humans are the data point on the extreme right.

to a number of core human activities, including lying and pretend play (Leslie 1987) and hence, in all likelihood, the human ability to create religious systems (most of which depend on the ability to imagine an alternative world, or that this world could be other than as it is if we could only prevail on some great force to make it so).

ToM has turned out to be an issue of considerable significance in developmental psychology because it seems that children are not born with this capacity but develop it soon after the age of 4 years. The speed with which they acquire it at this onto-genetic stage makes it unlikely that it is a purely learned effect, although it is likely that children use the preceding stage of "belief-desire psychology" to support the later emergence of ToM. More important, perhaps, there is no convincing evidence as yet to suggest that any species other than humans can aspire to the same ability. Even the tests carried out on chimpanzees have at best yielded equivocal results (O'Connell and Dunbar 2003; Call and Tomasello 1998).

This makes the true capacities of adult humans all the more impressive. Although there has been only one direct test of the full range of adult human abilities in this respect (Kinderman et al. 1998), it is clear from this study that normal adults can cope quite adequately with fifth-order intentionality ("I believe that you think that I want you to suppose that I believe that the world is flat"). Some individuals can cope with

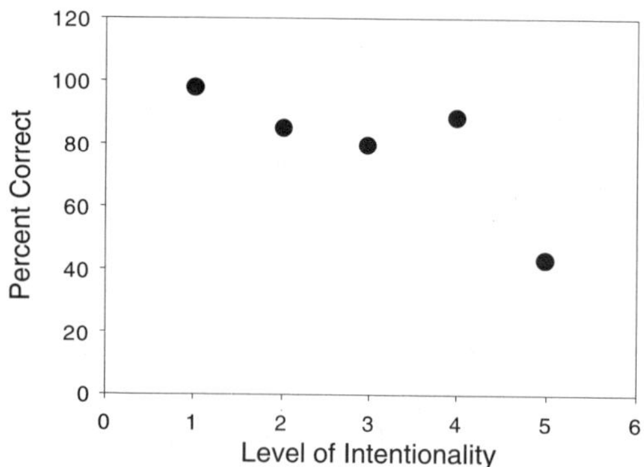

Figure 8.5

Performance of "advanced" theory-of-mind tasks by normal adult humans. (Redrawn from Kinderman et al. 1998.) Note that the levels of intentionality are all one unit higher than those originally suggested by Kinderman et al. (1998); this is because Kinderman et al. did not include the subjects' mind state in their analysis of the tasks.

sixth-order intentional statements, but success rates at sixth order are about half those at fifth order (figure 8.5). This makes the contrast between apes (who can barely master second-order tasks) and adult humans all the more striking. It seems, then, as though the computational demands of advanced ToM-type tasks are what we humans use our massive brains for.

While it is not clear just how we use our fifth-level ToM abilities in everyday life, it seems reasonable to conclude that the ability to think in depth about a problem may be beneficial in many other walks of life. This may be important in planning since planning involves mental rehearsal. We know very little about either of these features, although Byrne (1995) has repeatedly emphasized the importance of planning in the mental life of great apes (and humans). Mental rehearsal (meaning the ability to mull over a problem without engaging in hands-on experimentation) may also be crucial to planning (defined as the ability to think about and predict the outcomes of alternative scenarios for a particular event or circumstance).

Why Did a Theory of Mind Evolve?

These kinds of advanced cognitive abilities seem to be associated closely with imitation and teaching, since the ability to imitate a conspecific or teach another how to do something lies at the heart of cultural transmission. Quite how these components relate to each other, however, remains unclear. Nonetheless, the ability both to imitate and to teach may require advanced theory of mind. This is perhaps most obvious in respect to teaching.

To be able to teach another individual to perform an action effectively, the teacher must be able to understand why the student is making the mistakes he or she does. This is not to say that teaching cannot be done without theory of mind; it obviously can in the "Just copy what I'm doing" form. The issue I want to focus on has much more to do with teaching complex routines such as rituals (and their significance) or extended causal sequences (such as explanations). Teaching these kinds of phenomena efficiently depends on the teacher having a much clearer understanding of why the pupil is making mistakes (or even of predicting where and why a pupil might make mistakes) so as to be able to adjust the pace of learning and the way whatever has to be learned is presented to suit the individual's particular needs.

Although there have been a number of claims about animals teaching their offspring particular behaviors (nut-cracking in chimpanzees, how to kill in cats), it has to be said that none of these comes close to what we see in humans. Teaching is so ubiquitous at all levels of society that it surely sets the human species apart. Almost everything we do is acquired (or influenced) by teaching or imitation. Quite whether anything more than formal theory of mind (i.e., abilities beyond second-order intentionality) is needed for teaching remains far from clear, however. Does a teacher, for example, need to believe that the student supposes that the teacher wants the students to behave in a particular way for successful teaching to occur? We simply don't know because no one has really given it any thought.

Imitation is an even more difficult phenomenon in this respect. It certainly appears to be important in human cultural transmission, yet it is far from clear just what is involved. It is hard to see an obvious role for theory of mind (at any level) in the kinds of imitation that comparative psychologists worry about. Indeed, prior to the age at which they develop full theory of mind (roughly 4–5 years) young children are already inveterate imitators. Much of language acquisition, for example, may result from imitation rather than direct teaching. Young children seem so willing and able to absorb knowledge by imitation that they might be described simply as imitating machines.

However, the kinds of imitation seen in infants does not provide a complete description of all human imitation. Imitation by older individuals may be much more dependent on advanced theory of mind than seems to be the case for young children. There may be much more work involved in trying to figure out just what is going on, and why, when older individuals learn to imitate some cultural habit or meme. One plausible reason for this is that many of the things that adults learn in this way involve badges of group membership. Identifying the role that such behaviors play in advertising group membership and how they will be viewed by others, whether members of the group or strangers, may be a fundamental component of much cultural transmission. Belonging to a social group is so fundamental for humans that an intuitive understanding of how a particular style of behavior or dress functions to evoke a sense of group membership may be crucial. Making the wrong kind of style statement may be, literally, the kiss of death in traditional societies. Adults may therefore be a lot less automatic in their copying behavior than young children and engage instead in much more careful evaluation of the benefits and purpose of specific behaviors.

Thus although the cognitive mechanism of imitation itself may be quite primitive (in the sense that it does not require any kind of advanced social cognitive abilities), the use to which this phenomenon is put in adults may require much greater involvement of advanced ToM-like abilities.

This is perhaps a more important issue than one might imagine at first glance. Most of the attention in the literature on imitation has focused on the rather primitive forms of imitation seen in very young children, mainly because this is the most likely form to be found in animals. Hence, in their attempts to decide whether animals engage in imitation (and thus might possess true "culture"), comparative psychologists have emphasized the essentially mindless form of imitation that typifies the behavior of young children. As a consequence, most of the literature on cultural transmission in humans (from Boyd and Richerson 1984 to Blakemore 1999) has emphasized the fact that imitation can be mindless.

In one key respect, of course, this has been an appropriate stance to take. Modelers of cultural evolution have been keen to focus on something that is genuinely different from more conventional forms of selection. Learning by trial and error (itself probably a crucial element in human cultural transmission) has been regarded, quite correctly, as just another form of natural selection. Making cultural evolution genuinely interesting from an evolutionary point of view requires focusing on something that is radically different. Mindless copying that allows memes to evolve through Darwinian processes without being subject to the forces of natural selection imposed by the physical (or social) environment sets cultural evolution apart from more conventional Darwinian processes, and thus makes it evolutionarily interesting. However, this may overlook other processes of cultural transmission whose mechanisms may be more prosaic but whose impact on human social life may be correspondingly greater.

This brings us to what is perhaps the crucial issue; namely, just why are these processes so crucial for modern humans? What is it about life as a modern human (and I include our most recent ancestors possibly as far back as late *Homo erectus* under this rubric) that makes the trappings of culture—and thus the mechanisms on which it depends—so crucial?

The answer lies, I suspect, in that curious phenomenon of group membership to which I have already alluded. There can be no question about the fact that the most serious challenge to the fitness of individual humans lies in free riders. This problem has long been recognized in economics, where it surfaces as the "tragedy of the commons" or the common pool resources problem (Orstrom et al. 1994). In any social system where explicit or implicit cooperative agreements allow members to benefit from a communal solution to one of the problems of survival or reproduction, free riders who take the benefit without paying the cost always have an advantage (Enquist and Leimar 1990). Moreover, this advantage increases with social group size and the degree of its geographic dispersion. Free riders benefit from the ignorance (and thus naiveté) of their victims whenever the size of the community or its physical dispersion makes it difficult for community members to exchange information about the behavior of free riders. Free riding is thus likely to be an evolutionarily stable strategy (ESS) and evolve to fixation, irrespective of whether the strategy is genetically determined or learned. Since both large dispersed communities and communal solutions to the problems of survival are characteristic of modern humans, this problem is likely to be especially intrusive in our case irrespective of whether we view the human condition as that of the hunter-gatherer or the citizen of a postindustrial nation-state.

Sophisticated cognitive capacities that allow individuals to assess the intentions of potential interactees thus become of overriding importance for humans. Being able to figure out just what someone else is up to and whether their declarations of friendship and the willingness to engage in exchange relationships are honest is probably a crucial adaptation to survival in both traditional and postindustrial human societies (Dunbar 1999). Not being exploited by the hunter-gatherer equivalent of the used car salesman becomes an essential component of human fitness-maximizing strategies.

This problem has been alluded to by many evolutionary psychologists, usually in the form of the cheat detection module proposed by Cosmides and Tooby (1992). I am less convinced by the specifics of a cheat detection module than by the suggestion that deep cognitive processing abilities of the kind represented by ToM may be the key to this, perhaps with the application of these abilities to particular issues, such as detection of cheating, being learned by a combination of individual trial and error and the cultural transmission of rules of thumb (in the form of old wives' tales, proverbs, grandmother's wisdom).

The importance of displaying and reading group badges in this respect cannot be underestimated; it colors almost everything we do. I have elsewhere elaborated on the role that dialects may play in this process at the social level (Dunbar 1999; see also Nettle and Dunbar 1997, Nettle 1999). The importance of devices such as this is that they delineate in an easily recognizable (and to a large extent uncheatable) way those individuals with whom one can afford to engage in unrestricted exchanges on the assumption that the debt incurred will be paid back in due course, either via kin selection or directly because of the (learned) sense of obligation that exists among members of the same small social group. Mechanisms such as dialects and other more conventional forms of social badges (the design of clothing, hair styles, knowledge of particular rituals or origin stories) play a vital role in welding social groups together so that they function effectively as cooperative alliances. All of these are transmitted as cultural icons by imitation and/or social learning.

Conclusions

I have tried to suggest here that the evolution of culture is related to a particular problem that became especially intrusive during the later course of human evolution, namely, the problem of the free rider. The solution, I suggest, was to evolve the ability to think deeply about the behavior and intentions of other individuals in order to be able to detect and control free riders. This requires levels of intentionality well beyond those normally associated with social life in monkeys and apes. However, what may have made this quantum leap possible was a nonlinear relationship between visual processing demands and the size of the brain. After a crucial point represented by the body (brain) size of the great apes, further increases in brain size allowed a disproportionate volume of the available neocortical matter to be allocated to more specialized social cognition tasks rather than to visual processing.

References

Barton, R., and Dunbar, R. I. M. 1997. Evolution of the social brain. In A. Whiten and R. Byrne (eds.), *Machiavellian intelligence II* (pp. 240–263). Cambridge: Cambridge University Press.

Blakemore, S. 1999. *The meme machine.* Oxford: Oxford University Press.

Boyd, R., and Richerson, P. 1984. *Evolution and cultural selection.* Chicago: University of Chicago Press.

Byrne, R. 1995. *The thinking ape.* Oxford: Oxford University Press.

Call, J., and Tomasello, M. 1998. A nonverbal theory of mind test: The performance of children and apes. *Child Development* 70: 381–395.

Cosmides, L., and Tooby, J. 1992. Cognitive adaptations for social exchange. In J. H. Barkow, L. Cosmides, and J. Tooby (eds.), *The adapted mind* (pp. 163–228). Oxford: Oxford University Press.

Dunbar, R. I. M. 1992. Neocortex size as a constraint on group size in primates. *Journal of Human Evolution* 22: 469–493.

Dunbar, R. I. M. 1993. Coevolution of neocortex size, group size and language in humans. *Behavioral and Brain Sciences* 16: 681–735.

Dunbar, R. I. M. 1998. The social brain hypothesis. *Evolutionary Anthropology* 6: 178–190.

Dunbar, R. I. M. 1999. Culture, honesty and the freerider problem. In R. Dunbar, C. Knight, and C. Power (eds.), *The evolution of culture* (pp. 194–213). Edinburgh: Edinburgh University Press.

Enquist, M., and Leimar, O. 1990. The evolution of fatal fighting. *Animal Behavior* 39: 1–9.

Joffe, T., and Dunbar, R. I. M. 1997. Visual and socio-cognitive information processing in primate brain evolution. *Proceedings of the Royal Society of London Series B* 264: 1303–1307.

Kinderman, P., Dunbar, R. I. M., and Bentall, R. P. 1998. Theory-of-mind deficits and causal attributions. *British Journal of Psychology* 89: 191–204.

Kudo, H., and Dunbar, R. I. M. 2001. Neocortex size and social network size in primates. *Animal Behavior* 62: 711–722.

Leslie, A. M. 1987. Pretence and representation in infancy: The origins of "theory of mind." *Psychological Review* 94: 84–106.

Lewis, K. 2000. A comparative study of primate play behaviour: Implications for the study of cognition. *Folia Primatologica* 71: 417–421.

Nettle, D. 1999. *Linguistic diversity*. Oxford: Oxford University Press.

Nettle, D., and Dunbar, R. I. M. 1997. Social markers and the evolution of reciprocal exchange. *Current Anthropology* 38(1): 93–99.

O'Connell, S., and Dunbar, R. I. M. 2003. A test for comprehension of false belief in chimpanzees. *Evolution & Cognition* 9: 131–139.

Orstrom, E., Gardner, R., and Walker, J. 1994. *Rules, games and common-pool resources*. Ann Arbor, Mich.: University of Michigan Press.

Stephan, H., Frahm, H. D., and Baron, G. 1981. New and revised data on volumes of brain structures in insectivores and primates. *Folia Primatologica* 35: 1–29.

Pawlowski, B. P., Lowen, C. B., and Dunbar, R. I. M. 1998. Neocortex size, social skills and mating success in primates. *Behavior* 135: 357–368.

Whiten, A., Goodall, J., McGrew, W. C., Nishida, T., Reynolds, V., Sugiyama, Y., Tutin, C. E. G. Wrangham, R. C., and Boesch, C. 1999. Cultures in chimpanzees. *Nature* 399: 682–685.

9 The Evolution of Culture from a Neurobiological Perspective

Wolf Singer

The emergence of culture is in all likelihood the result of synergistic interactions among several evolutionary events. First, the development of bipedal gait freed the arms from locomotor functions and permitted refinement of the hands and ultimately the sophisticated use of tools. Second, the formation of labor-sharing societies freed individuals from devoting all their time to ensuring survival and reproduction and permitted a more relaxed schedule of occupations. Third, the invention of agriculture and cattle breeding allowed a sedentary lifestyle and hence a more continuous investment in technological advances. Fourth, the development of language as a symbolic communication system permitted easy transfer of acquired knowledge among individuals of the same generation as well as across generations. However, none of these achievements would have been possible without the evolutionary refinement of brain functions. The use of tools, the formation of labor-sharing societies, the management of agriculture, and the development of language are consequences of the increasing sophistication of the cognitive functions and motor skills provided by the brain of *Homo sapiens sapiens*. Because our direct ancestors are all extinct, it is extremely difficult to infer which aspects of brain development have actually been decisive for the transition from apes to early hominids and finally culture-competent *Homo sapiens sapiens*. The only truly longitudinal data on the evolution of the human brain come from studies of fossil skulls. These analyses reveal a gradual increase in brain volume, but this notion is not particularly helpful because brain size alone, even if considered in relation to body weight, is only a poor indicator of functional sophistication. Thus, inferences about evolutionary changes in the organization of the brain have to rely on comparisons of species that escaped extinction. However, and this is both interesting and unfortunate for studies of evolution, the surviving species all branched off from the line of our ancestors long before the gap that separates us from our nearest relatives, the apes. Therefore, only rather indirect inferences are possible.

What Makes the Difference?

The first question is whether our brains differ from those of our ancestors who initiated cultural evolution, painted the walls of their caves, and invented tools. The answer is "yes and no." As far as genetically determined features are concerned, i.e., the molecular composition, the anatomical structures, and the basic connectivity patterns, there cannot be any major differences because evolution is slow. This implies that our ancestors were born with brains that must have had the same basic abilities as those of modern babies. Hence, competencies exceeding those of modern cave-dwelling ancestors must be attributed to the action of epigenetic factors, i.e., to experience-dependent modifications of postnatal brain development, and to learning.

At the neuronal level, these two processes of knowledge acquisition differ mainly with respect to the reversibility and the amplitude of the changes. Both are associated with modifications of the interactions among neurons. During early development, experience can modify the architecture of neuronal connectivity by gating the consolidation and disruption of newly formed pathways. Once these developmental processes decline, which is thought to occur around puberty, further modifications of functional architectures appear to be restricted to changes in the efficacy of the now-consolidated repertoire of connections (for a review see Singer 1990, 1995). Thus, although the genetically determined blueprint of our brains is probably the same as that of our ancestors, our brains are likely to differ because of differences in epigenetic shaping of fine structure.

This susceptibility of brain functions to epigenetic modifications is certainly a relevant factor in cultural evolution because it permits the highly efficient transmission of acquired abilities and knowledge from one generation to the next. However, the cognitive abilities acquired through this epigenetic path are a consequence of cultural evolution and collective learning rather than its cause. What, then, could have been the decisive steps in the evolution of the brain that actually triggered the onset of cultural evolution?

Triggers of Cultural Evolution

Concerning specific sensory or motor skills, *Homo sapiens sapiens* is a generalist. We perform reasonably well in many domains, but for most of them one can identify animals that outperform us. Still, if one distributed points for performance in the various sensory modalities and for basic motor skills, we would with all likelihood come out as winners. It is unlikely, however, that this superiority in average performance is alone sufficient to account for the emergence of culture. Rather, our cultural competence seems to result from the evolutionary development of certain cognitive functions that are unique to humans. One of these is probably our ability to generate abstract, symbolic metarepresentations of cognitive contents by subjecting the results of first-order cognitive operations iteratively to further cognitive processing of a higher order. This competence requires the ability to bind the results of distributed cognitive processes together and to re-represent the results of these binding operations at higher processing levels. The results of such higher-order cognitive operations are modality invariant and hence they are abstract descriptions of the outcome of first-order cognitive operations. Since these higher-order descriptions are equivalent with an internal protocol that keeps track of the brain's own cognitive operations, they can be considered as the substrate for our ability to be aware of our own sensations and intentions as well as those of others. This awareness in turn is probably the origin of our unique ability to use a symbolic communication system and to generate a theory of mind.

We seem to be the only species that is capable of imagining the thoughts and emotions of another person in a particular situation. We are the only species capable of entering into dialogues of the form, "I know that you know that I know," or "I know that you know how I feel." Such dialogues not only permit a deeper understanding of the respective other but they also allow one to experience one's own cognitive functions in the reflection of the perceptions of the other. Thus, the ability to generate a theory of mind has probably also been instrumental in shaping concepts of self, in creating autonomous agents endowed with intentionality and free will, and in the creation of social realities such as value systems. These cultural constructs are as real as the precultural realities. They are part of the environment in which human beings evolve and hence are likely to have as important a role in the epigenetic shaping of the brain's functional architecture as the other environmental factors. This has profound consequences. If one accepts that the fine-grained architecture of brains is part of the phenotype of an organism, then cultural embedding influences the phenotype, and because it does so as a function of experience gathered by preceding generations, cultural evolution follows principles that are clearly different from those ruling biological evolution.

Neuronal Prerequisites

What do we know about the neuronal substrate that enables human brains to generate metarepresentations of their own processes, to realize what one might call an inner eye function, and to develop a theory of mind? What permits our brains to evaluate and represent relationships among the distributed results of basic cognitive operations that occur in parallel and in relative isolation at lower levels of the brain? What could be the structure of the resulting metarepresentations that transcend the modality-specific descriptions provided by the sensory systems and that permit symbolic encoding of both external events and internal states?

One prerequisite for the generation of such higher-order descriptions and for their translation into behavior is a mechanism that binds the results of distributed first-order processes and makes the contents of these higher-order representations available to executive systems. At the neuronal level, the following requirements have to be fulfilled: (1) All computational results, both those of first- and of higher-order processes, must be expressed in a common format to permit flexible recombination. (2) A versatile binding mechanism must be implemented that permits evaluation and representation of the relationships among the results of distributed computations. (3) Expanded storage capacities must be provided to maintain temporally dispersed contents in short-term buffers so that they are simultaneously available for the evaluation of relations and for binding. (4) Sufficient neuronal substrate needs to be provided for the generation of higher-order descriptions. (5) Effector systems are required that are sufficiently differentiated to permit the translation of the results of higher-order computations into actions and the communication of the symbolic descriptions to other individuals.

When comparing our brains with those of nonhuman primates, one is struck by their similarity and searches in vain for entirely novel structures that could account for the new, qualitatively different functions. At the macroscopic level, the only noticeable difference between our brains and those of nonhuman primates is an increase in the surface of the neocortex, and these differences vanish nearly completely if one analyzes the brains at the microscopic or molecular level. The internal organization of the brain structures, including the neocortex, is nearly identical, and the vast majority of the molecules expressed are the same. This leaves one with the conclusion that the new functions that distinguish *Homo sapiens sapiens* from its nearest neighbor species must have been realized simply by the addition of further areas of the neocortex and/or by the rearrangement of connections among neocortical areas.

Comparative anatomy suggests that these additional areas differ from the more ancient areas in the way in which they are connected to sensory systems and effector organs. The new areas in the occipital, parietal, and temporal lobes appear to be involved primarily in the refinement of sensory functions, whereas the new areas in the frontal lobes subserve more executive functions, such as planning action, short-term storage, and management of attention. The more recent sensory areas tend to receive their input, not directly from the sensory periphery, as is the case for the more ancient sensory areas, but more indirectly via the latter. Moreover, the new areas tend to collect their input, not from a single modality as the ancient sensory areas do, but from different modalities (Krubitzer 1995, 1998). It is because of this peculiarity that the phylogenetically more recent areas that are topographically intercalated between the monomodal sensory areas have been viewed as association areas. The new areas in the frontal lobes are also more remote from the periphery than the more ancient motor centers. They tend to be connected to effector organs only indirectly via the ancient motor areas and receive most of their cortical input, not from the primary sensory areas, but from the more recent association areas. Thus, as indicated in figure 9.1, the connectivity of these phylogenetically recent cortical areas is compatible with the view that they reevaluate and bind the distributed results of primary cognitive operations and thereby provide the substrate for the generation of higher-order representations.

What remains puzzling, however, is the fact that all these different functions appear to always rely on the same computational algorithm. The intrinsic organization of the neocortex is extremely complex but surprisingly stereotyped and monotonous. The laminar organization, the various cell types, and the intrinsic connectivity differ only little between phylogenetically old and more recent cortical areas or between areas devoted to sensory and those devoted to executive functions. Because the program for the computational operations performed by neuronal networks is fully and exclusively determined by the architecture and coupling strength of connections, the structural homogeneity of the neocortex implies that the various regions perform more or less the same computations. This fascinating conclusion has recently received strong support from developmental studies in which inputs from the eye have been rerouted by surgical intervention to the auditory cortex, whereupon this piece of cortex developed exactly the same functional features that are normally characteristic for the visual cortex (Sharma et al. 2000).

It appears then as if our brains owe their unique cognitive abilities simply to the iteration of processes realized by cortical architectures. All that seemed necessary for the development of new functions was apparently the addition of cortical areas that treat the output of the already existing areas in exactly the same way as these treat their input, which in lower animals comes mainly from the sensory periphery. In conclusion, the new cognitive abilities that distinguish humans from nonhuman primates seem to have emerged because evolution provided additional cortical areas that permitted reprocessing and binding of the results of first-order processes and the generation of higher-order, transmodal representations. It is interesting that the evolution of new cortical areas may not have required major changes in the genome, because adding one more step of cell division to the division cycles of the precursor cells of neocortical neurons can have dramatic effects on cortical cell numbers and hence cortical volume (Rakiç 1998).

The notion that neocortical modules process signals according to similar algorithms has the additional attractive implication that the results of their computations are likely to be encoded in the same format. Hence, they can be resubjected in ever-changing constellations to iterative processes of the same kind, thus generating representations of increasingly higher order. Although this view is far from providing a mechanistic explanation for the emergence of capacities such as phenomenal awareness, i.e., the ability to be aware of one's own sensations and actions, it provides at least an intuition of how brains can apply their cognitive abilities to some of their

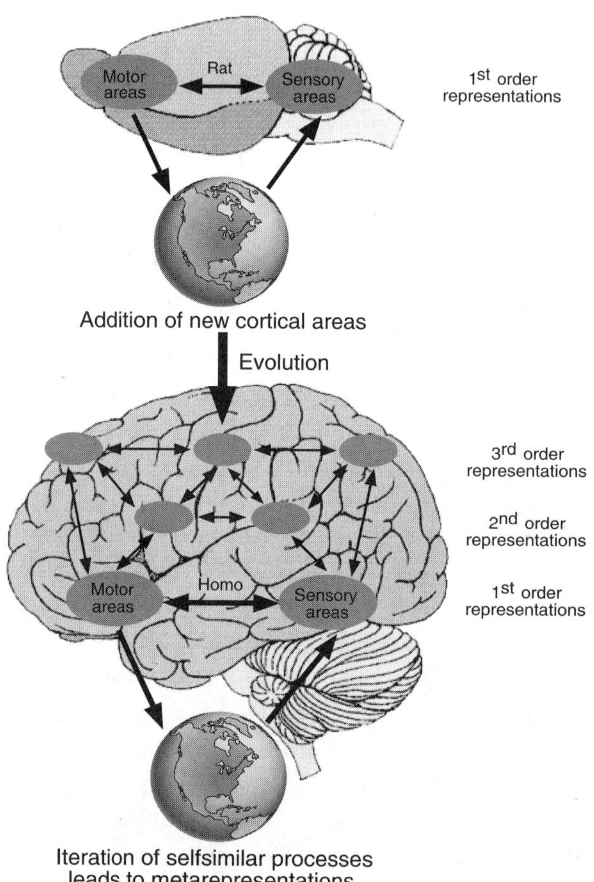

Figure 9.1

Emergence of phenomenal awareness. The evolution of the mammalian brain is characterized by an increase in the surface of the cerebral cortex and the addition of new cortical areas (for further comments, see text).

own processes, thereby creating descriptions of themselves and hypotheses about others.

In conclusion, it appears that any attempt to account for the emergence of those cognitive abilities that we consider instrumental for the evolution of culture needs to be based on an understanding of neocortical functions, and in particular those that permit the binding of the results of distributed, primary cognitive operations into coherent, modality-independent, and symbolic descriptions. Of primary interest, therefore, is how contents are represented in cortical networks and how dynamic binding of these contents into metarepresentations can be achieved. In the following section these questions are investigated using visual perception and the organization of the mammalian visual system as an example.

The Structure of Representations

The hypothesis proposed here is that evolved brains use two complementary strategies to represent contents (see also Singer 1995, 1999). The first strategy relies on individual neurons that are tuned to respond selectively to particular constellations of input activity, thereby establishing explicit representations of particular constellations of features. It is commonly held that the specificity of these neurons is brought about by selective convergence of input connections in hierarchically structured feedforward architectures (figure 9.2). This representational strategy allows rapid processing and is ideally suited for the representation of frequently occurring stereotyped combinations of features. However, this strategy has several limitations. It is expensive in terms of the number of neurons required because it demands at least one neuron per object. Thus it is not well suited to cope with the virtually infinite diversity of possible constellations of features encountered in real-world objects. Moreover, this representational mode lacks systematicity, which makes it difficult to encode relations between parts of the same object or semantic relations between different perceptual objects. A detailed discussion of the advantages and disadvantages of representing contents by individual smart neurons is to be found in Singer (1999), von der Malsburg (1999), and Gray (1999).

The second strategy consists of the temporary association of neurons into a functionally coherent assembly that as a whole represents a particular content by which each of the participating neurons is tuned to one of the elementary features of the perceptual object (figure 9.3). This representational strategy is more economical with respect to neuron numbers than the first one because a particular neuron can at different times participate in different assemblies, just as a particular feature can be shared by many different perceptual objects. Moreover, this representational strategy allows the rapid de novo representation of constellations of features that have never been experienced before. There are virtually no limits to the dynamic association of neurons in ever-changing constellations, provided that the participating neurons are directly or

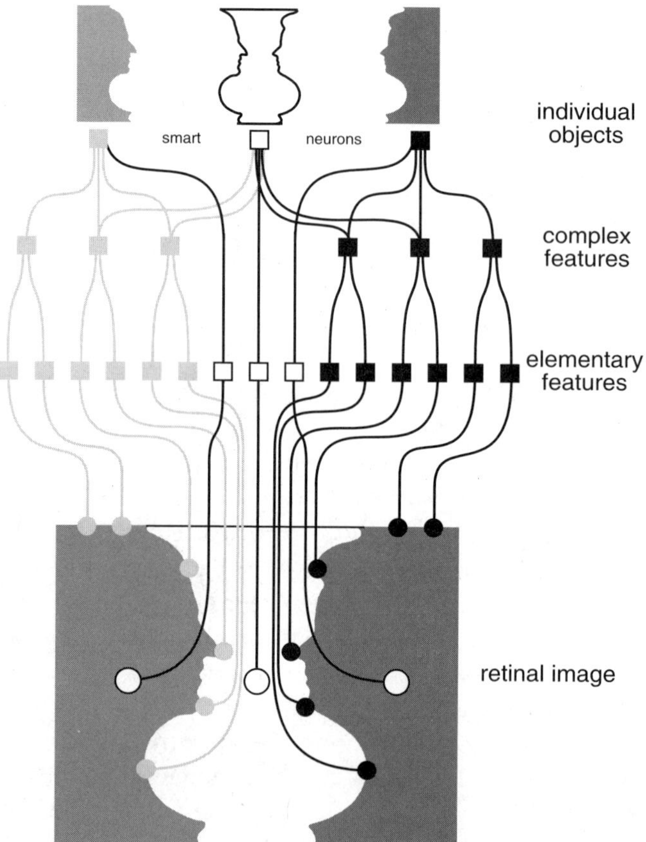

Figure 9.2

The representation of perceptual objects by smart neurons. It is assumed that the distributed elementary features of perceptual objects are bound by convergence in hierarchically structured feedforward architectures, leading to highly specialized neurons at the top of the processing stream that respond selectively to the constellation of features characterizing individual objects.

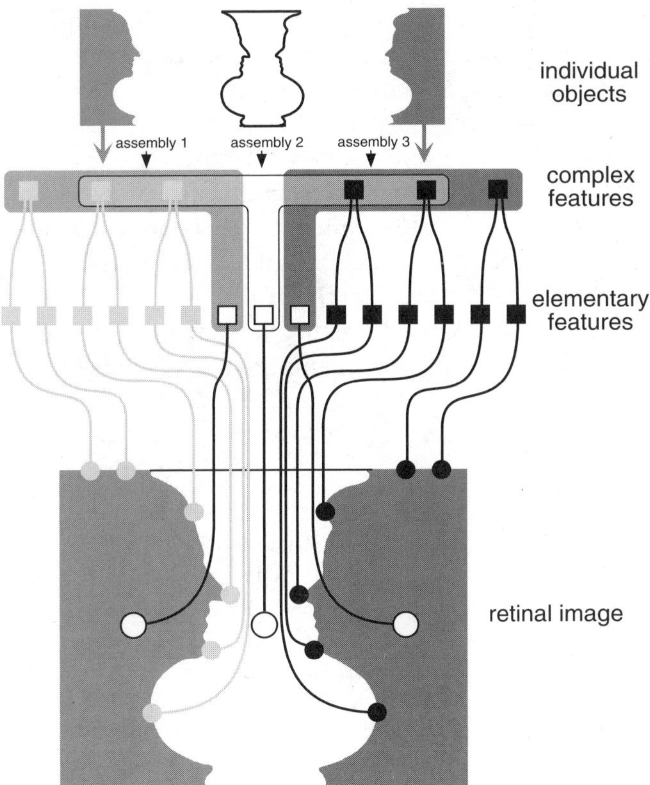

Figure 9.3

The representation of perceptual objects by dynamically associated assemblies of neurons. It is assumed that the specific constellation of features characterizing a perceptual object is represented by an assembly of distributed neurons, each of which is tuned to a moderately complex constellation of elementary features. These assemblies stabilize through reciprocal cooperative interactions that are mediated by a dense network of association connections whose functional architecture is modifiable by learning. Note that the same set of feature-selective neurons can participate in the representation of different objects. Cells responding to the contours of the vase can be bound either in a single assembly representing the vase (assembly 2) or in two different assemblies representing the two faces (assemblies 1 and 3).

indirectly connected. Thus, for the representation of highly complex and permanently changing contents, this second strategy appears to be better suited than the first.

The metarepresentations that result from the iteration of cognitive operations are necessarily much richer in combinatorial complexity than the contents of first-order processes. In addition, they must be highly dynamic because they need to be reconfigured at the same pace as the changes in contents of phenomenal awareness. It appears then as if the second representational strategy, which is based on the dynamic binding of neurons into functionally coherent assemblies, would be more suitable for the implementation of higher-order representations than the first strategy, which relies on individual smart neurons. While the latter can be readily implemented in simple feedforward networks and hence can be found also in the brains of invertebrates, assembly coding requires neuronal architectures that permit, in addition, dynamic grouping of distributed responses through reentry and self-organization. This necessitates cooperative interactions among neurons and hence a complex network of reciprocal connections. It appears as if such architectures existed only in cortical structures, which may be one reason for the evolutionary success of the cerebral cortex.

The following sections, therefore, focus on the question of whether there is any evidence that contents are represented in the cerebral cortex, not only explicitly by smart neurons that are tuned to represent highly complex constellations of features, but also by dynamically associated assemblies of distributed neurons.

The Signature of Assemblies

In assembly coding, two important conditions need to be met. First, a selection mechanism is required that permits a dynamic yet consistent association of neurons into distinct, functionally coherent assemblies. Second, responses of neurons that have been identified as groupable must be labeled so that they can be recognized in subsequent processing as belonging together. This is necessary to ensure that responses, once they are bound together, are evaluated jointly as constituents of a coherent code and are not confounded with the responses of cells belonging to other, simultaneously formed assemblies that represent different contents. Following the initial proposal of assembly coding by Hebb (1949), numerous theoretical studies have addressed the question of how assemblies can self-organize on the basis of cooperative interactions within associative neuronal networks (for a review, see Philipps and Singer 1997). Here I focus on the second problem of assembly coding, the question of how responses of cells that have been grouped into an assembly can be tagged as related. An unambiguous signature of relatedness is absolutely crucial for assembly codes because, unlike in explicit single-cell codes, the meaning of responses changes with the context in which they are interpreted. Hence, in assembly coding, false conjunctions are deleterious.

Tagging responses as related is equivalent to ensuring that they are processed and evaluated together at the subsequent processing stage. This in turn can only be achieved by raising their salience jointly and selectively, and there are three options. First, nongrouped responses can be inhibited; second, the amplitude of the selected responses can be enhanced; and third, the selected cells can be made to discharge in precise temporal synchrony. All three mechanisms enhance the relative impact of the grouped responses at the next higher processing level. Selecting responses by modulating discharge rates is not problematic in processes that rely on smart neurons because a particular cell always signals the same content. However, selecting responses by raising discharge rates may not always be suitable for the distinction of assemblies because it introduces ambiguities (von der Malsburg and Schneider 1986) and reduces processing speed (Singer et al. 1997). Ambiguities could arise because the discharge rates of feature-selective cells vary over a wide range as a function of the match between stimulus and receptive field properties, and these modulations in response amplitude would not be distinguishable from those signaling the relatedness of responses. Processing speed would be reduced because it takes some time to find out which neurons discharge at high and low rates, respectively. This distinction can only be made once a sufficient number of successively arriving synaptic potentials have been integrated by cells at the next processing stage. Thus, rate-coded assemblies need to be maintained for some time in order to be distinguishable. However, different assemblies cannot overlap in time within the same processing stage; first, because it would be impossible to distinguish which responses belong to which assembly, and second, because different assemblies might have to share the same neurons. Thus, if assemblies were solely distinguished by enhanced firing of the participating neurons, ambiguities could arise on a result of rate fluctuations that are unrelated to binding functions. Moreover, the rate at which different contents could be successively represented by a given population of neurons would be slow.

Both limitations can be overcome if the selection and labeling of responses is achieved through synchronization of individual discharges, i.e., by establishing precise temporal relations among the discharge probabilities of distributed neurons (von der Malsburg and Schneider 1986; Gray et al. 1989). Expressing the relatedness of responses by synchronization resolves the ambiguities resulting from stimulus-dependent rate fluctuations because synchronization of firing on a short time scale of a few milliseconds can be modulated independently of rate fluctuations occurring at time scales of tens or hundreds of milliseconds. Slowly varying response amplitudes could thus be reserved to signal how well particular features match the preferences of neurons, and synchronicity could be used in parallel to signal how these features are related. Defining assemblies by synchronization also accelerates the rate at which different assemblies can follow one another because the selected event is the individual spike or a brief burst of spikes; salience is enhanced only for those discharges that are precisely synchronized and generate coincident synaptic potentials in target cells at the subsequent processing stage. The rate at which different assemblies can follow one another without becoming confounded is then limited only by the duration of the interval over which synaptic potentials summate effectively, and this can be as short as 10 ms or less (for a detailed discussion of binding by synchrony, see Singer 1999).

Another advantage of selecting responses by synchronization is that the timing of input events is preserved with high precision in the output activity of cells because synchronized input is transmitted with minimal latency jitter (Diesman et al. 1999). This in turn can be exploited to preserve the signature of relatedness across successive processing steps, which is of great advantage if relationships of a higher order need to be encoded.

Behavioral Correlates of Response Synchronization

Following the discovery of stimulus-related response synchronization among neurons in the cat visual cortex (Gray and Singer 1987, 1989), numerous experiments have been performed to find out whether synchronization of neuronal activity is at all of functional significance and in particular whether it could serve as signature of relatedness in assembly coding and hence be used as a general binding mechanism for the generation of representations. One of the predictions to be tested was that synchronization probability should reflect some of the gestalt criteria according to which the visual system groups related features during scene segmentation. Among the grouping criteria examined so far are continuity, vicinity, similarity, and colinearity in the orientation domain, and common fate in the motion domain (Gray et al. 1989; Engel et al. 1991a,b; Freiwald et al. 1995; Castelo-Branco et al. 2000 for the cat; Kreiter and Singer 1996 for the monkey). The results of these investigations are fully compatible with the hypothesis that the probability of response synchronization reflects the gestalt criteria applied for perceptual grouping. Most important, none of these synchronization phenomena were detectable among responses that were evoked by the same stimulus but recorded successively rather than simultaneously. This indicates that synchronization was not simply due to stimulus-induced covariations in discharge rate, but resulted from internal dynamic coordination of spike timing. The temporal coherence observed among simultaneously recorded responses was much greater than that expected from mere covariation of event-related fluctuations in the discharge rate.

The interactions responsible for these dynamic synchronization phenomena are mediated to a substantial extent by connections that reciprocally interconnect cortical neurons (Engel et al. 1991c; Nowak et al. 1995; Löwel and Singer 1992; König et al. 1993). The architecture of these connections agrees with the hypothesis that they serve to group responses according to common gestalt criteria. They preferentially link neurons that are tuned to features that tend to be grouped perceptually (Das and Gilbert 1999; Schmidt et al. 1997).

Synchronization and Attention

The perception of complex relations among features of objects, in particular when these are encoded in different modalities, usually requires that one direct one's attention to the task (Treisman and Gelade 1980). If synchronization serves as a code for the definition of higher-order relations in cognitive processes, one expects that it will be particularly prominent when the brain is aroused and engaged in tasks requiring attentive scrutinizing and binding of features. Evidence confirming this expectation has been obtained. Content-specific response synchronization occurs only when the brain is in an activated state (Munk et al. 1996; Herculano et al. 1999) and increases further when the animals attention is focused on a particular cognitive task (Roelfsema et al. 1997; Steinmetz et al. 2000; Fries et al. 2001). Related findings have been obtained in human subjects. Here it is not possible to directly measure the synchronization of individual neurons, but synchronization of cell populations can be seen in the electroencephalogram. The reason is that synchronization tends to be associated with a rhythmic patterning of the responses in the range of 40 Hz, the so-called gamma-frequency range (Gray and Singer 1989), and these oscillations are readily distinguishable in recordings from the scalp. These investigations have revealed an increase in gamma oscillations and enhanced phase locking of oscillatory activity when subjects were involved in cognitive tasks that required selective attention (Tiitinen et al. 1993), feature binding (Tallon-Baudry et al. 1997; Rodriguez et al. 1999; Keil et al. 1999), storage in short-term memory (Tallon-Baudry et al. 1998), and associative learning (Miltner et al. 1999).

Close correlations have also been found between response synchronization and changes in perception that result from internal, self-generated switches in response selection. One of these examples is binocular rivalry. When the two eyes are presented with patterns that cannot be fused into a single coherent percept, the two patterns are perceived in alternation rather than as a superimposition of their components. This implies that there is a central gating mechanism that selects in alternation the signals arriving from the two eyes for further processing. Interocular rivalry is thus a suitable paradigm for investigating the neuronal correlates of dynamic response selection and binding.

This paradigm has been used to investigate how neuronal responses that are selected and perceived differ from those that are suppressed and excluded from supporting perception (Fries et al. 1997, 2002). Because the animal performs tracking eye movements only for the pattern that is actually perceived, patterns moving in opposite directions were presented dichoptically in order to determine from the optokinetic tracking response which of the two eyes was selected. These experiments revealed close correlations between changes in the strength of response synchronization and the outcome of rivalry. Cells mediating responses of the eye that won in interocular competition increased the synchronicity of their responses whereas the reverse was true for cells driven by the eye that became suppressed. Thus, in agreement with the hypothesis, selection of responses for further processing appears to be achieved by raising their salience through synchronization. Likewise, suppression is not achieved by inhibiting responses, but by desynchronization (figure 9.4).

A particularly close correlation between neuronal synchrony and perceptual grouping has been observed in experiments with plaid stimuli. These stimuli are well suited for the study of dynamic binding mechanisms because minor changes in the stimulus cause a binary switch in perceptual grouping. Two superimposed gratings moving in different directions (plaid stimuli) may be perceived either as two surfaces, one being transparent and sliding on top of the other (component motion), or as a single surface consisting of crossed bars, which moves in a direction intermediate to the component vectors (pattern motion) (Adelson and Movshon 1982; Stoner et al. 1990). Which percept dominates depends on the luminance of the grating intersections because this variable defines the degree of transparency (Albright and Stoner 1995). Component (or pattern) motion is perceived when luminance conditions are compatible (or incompatible) with transparency (figure 9.5A). In the case of component motion, the responses evoked by the two gratings must be segregated, and only responses evoked by the contours of the same grating must be grouped to represent one of the two surfaces. In the case of pattern motion, responses to all contours must be bound together to represent a single surface. If this grouping of responses is initiated by selective synchronization, three predictions must hold (see figure 9.5B). First, neurons that prefer the direction of motion of one of the two gratings and have colinearly aligned receptive fields should always synchronize their responses because they always respond to contours that belong to the same surface. Second, two neurons that are tuned to the motion directions of the two gratings should synchronize their responses in the case of pattern motion because they then respond to contours of the same surface, but they should not synchronize in the case of component motion because their responses are then evoked by contours belonging to different surfaces. Third, neurons preferring the direction of pattern motion should also synchronize only in the pattern and not in the component motion condition.

Cross-correlation analysis of responses from cell pairs distributed either within or across areas of the visual cortex confirmed all three predictions. Cells synchronized their activity if they responded to contours that were perceived as belonging to the same surface (Castelo-Branco et al. 2000) (see figure 9.5C).

These results provide direct support for the hypothesis that precise temporal relations between the discharges of spatially distributed neurons play an important role in cortical processing and that synchronization is exploited to bind responses for

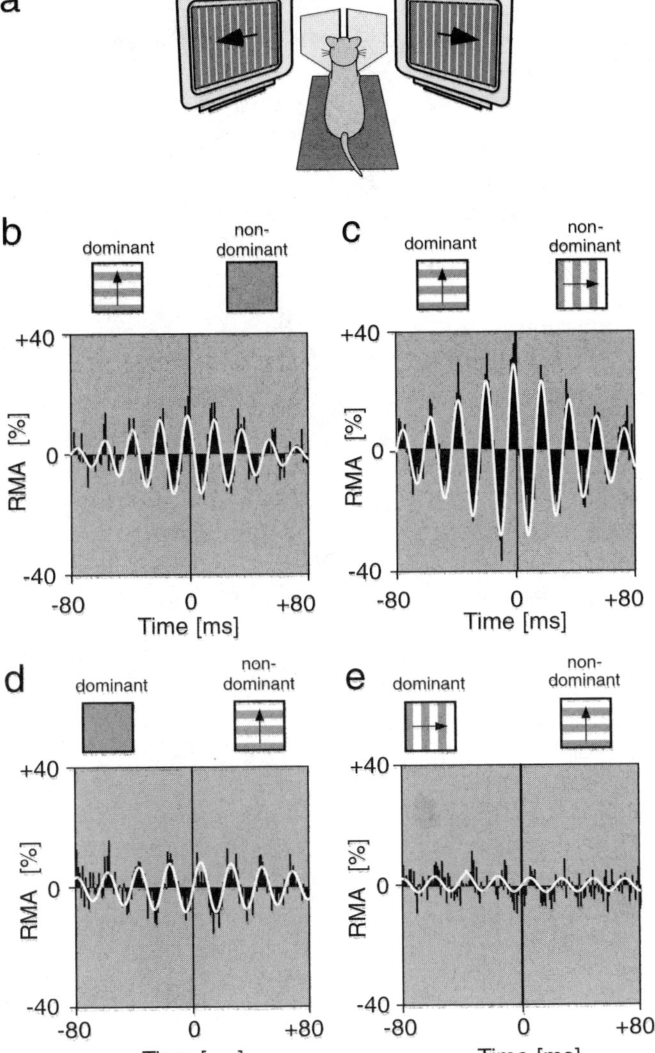

Figure 9.4

Neuronal synchronization under conditions of binocular rivalry. (a) Using two mirrors, different patterns were presented to the two eyes of strabismic cats. Panels (b)–(e) show normalized cross-correlograms for two pairs of recording sites activated by the eye that won (b,c) and lost (d,e) in interocular competition, respectively. The insets above the correlograms indicate stimulation conditions. Under monocular stimulation (b), cells driven by the winning eye show a significant correlation that is enhanced after introduction of the rival stimulus to the other eye (c). The reverse is the case for cells driven by the losing eye (compare conditions d and e). The white continuous line superimposed on the correlograms represents a damped cosine function fitted to the data. RMA is the relative modulation amplitude of the center peak in the correlogram, computed as the ratio of peak amplitude over the offset of correlogram modulation. This measure reflects the strength of synchrony. (Modified from Fries et al. 1997.)

A

B

Predictions on Synchronization

Stimulus configuration	Predictions
component:	sync
pattern:	sync
component:	no sync
pattern:	sync
component:	no sync
pattern:	sync

C

further joint processing by raising their salience. Thus, synchronization fulfills the requirements postulated for a binding mechanism that selects a subset from a larger number of simultaneously active neurons and labels the responses of this subset in a way that favors joint treatment during subsequent processing. By iteration of this binding operation, higher-order representations can be generated that remain distributed and contain nonlocal descriptions of internal states (for a detailed discussion of this hypothesis and a comprehensive review of the extensive literature on this issue, see Singer 1993, 1999; Singer and Gray 1995; Engel and Singer 2001).

Conclusion and Outlook

In summary, the hypothesis defended here is based on the following assumptions: (1) Cultural evolution has become possible mainly because of the evolution of cognitive abilities that permit the development of a theory of mind. (2) These abilities result from processing architectures that allow the generation of metarepresentations of brain states. (3) The neuronal substrate for these operations is the cerebral cortex, and the decisive evolutionary step that allowed the generation of metarepresentations was the addition of new cortical areas that process the output of more ancient areas in the same way as these process their respective sensory inputs. (4) These metarepresentations consist of dynamically configured cell assemblies and require for their generation the dynamic binding of distributed neurons into functionally coherent groups. (5) The binding mechanism that groups neurons into such assemblies and labels their responses as related is the transient synchronization of discharges.

Because these computations are similar in all cortical areas, the respective results are always available in the same format. Together with a flexible binding mechanism, this provides the option for virtually infinite permutations in the recombination of

◀ **Figure 9.5**

(A) Two superimposed gratings that differ in orientation and drift in different directions are perceived either as two independently moving gratings (component motion) or as a single pattern drifting in the intermediate direction (pattern motion), depending on whether the luminance conditions at the intersections are compatible with transparency. (B) Predictions of the synchronization behavior of neurons as a function of their receptive field configuration (left) and stimulation conditions (right). (C) Changes in synchronization behavior of two neurons recorded simultaneously from areas 18 and PMLS that were activated with a plaid stimulus under component and pattern motion conditions. The two neurons preferred gratings with orthogonal orientation (see receptive field configuration, top, and tuning curves obtained with component and pattern, respectively) and synchronized their responses only when they were activated with the pattern stimulus (compare cross-correlograms on the right). (Adapted from Castelo-Branco et al. 2000.)

distributed computational results. This raises the question of where the limits of such a system are and whether it could be evolved further. One variable limiting the combinatorial space is the finite dynamic range of neurons. It restricts the number of inputs that can be integrated and distinguished by an individual neuron, and hence the number of different constellations into which a given neuron can be bound. This constraint to some extent can be overcome by distributed processing on the one hand and by the creation of symbolic representations on the other. In the evolution of mammalian brains, both options have been exploited. There is a massive increase of functionally specialized cortical areas that support distributed processing. At the same time, these new areas have been embedded in architectures that permit the generation of higher-order representations and symbolic codes through dynamic binding. All these computational principles are realized in the brains of mammals and nonhuman primates. The evolutionary changes in brain architectures were gradual and quantitative rather than qualitative, and it is at least worth wondering whether there is a natural limit to the evolution of complex brains and whether it has been reached in *Homo sapiens*.

References

Adelson, E. H., and Movshon, J. A. 1982. Phenomenal coherence of moving visual patterns. *Nature* 300: 523–525.

Albright, T. D., and Stoner, G. R. 1995. Visual motion perception. *Proceedings of the National Academy of Sciences of the USA* 92: 2433–2440.

Castelo-Branco, M., Goebel, R., Neuenschwander, S., and Singer, W. 2000. Neural synchrony correlates with surface segregation rules. *Nature* 405: 685–689.

Das, A., and Gilbert, C. D. 1999. Topography of contextual modulations mediated by short-range interactions in primary visual cortex. *Nature* 399: 655–661.

Diesmann, M., Gewaltig, M. O., and Aertsen, A. 1999. Stable propagation of synchronous spiking in cortical neural networks. *Nature* 402: 529–533.

Engel, A. K., Kreiter, A. K., König, P., and Singer, W. 1991a. Synchronization of oscillatory neuronal responses between striate and extrastriate visual cortical areas of the cat. *Proceedings of the National Academy of Sciences of the USA* 88: 6048–6052.

Engel, A. K., König, P., and Singer, W. 1991b. Direct physiological evidence for scene segmentation by temporal coding. *Proceedings of the National Academy of Sciences of the USA* 88: 9136–9140.

Engel, A. K., König, P., Kreiter, A. K., and Singer, W. 1991c. Interhemispheric synchronization of oscillatory neuronal responses in cat visual cortex. *Science* 252: 1177–1179.

Engel, A. K., and Singer, W. 2001. Temporal binding and the neural correlates of sensory awareness. *Trends in Cognitive Sciences* 5: 16–25.

Freiwald, W. A., Kreiter, A. K., and Singer, W. 1995. Stimulus-dependent intercolumnar synchronization of single-unit responses in cat area 17. *Neuroreport* 6: 2348–2352.

Fries, P., Roelfsema, P. R., Engel, A. K., König, P., and Singer, W. 1997. Synchronization of oscillatory responses in visual cortex correlates with perception in interocular rivalry. *Proceedings of the National Academy of Sciences of the USA* 94: 12699–12704.

Fries, P., Reynolds, J. H., Rorie, A. E., and Desimone, R. 2001. Modulation of oscillatory neuronal synchronization by selective visual attention. *Science* 291: 1560–1563.

Fries, P., Schröder, J.-H., Roelfsema, P. R., Singer, W., and Engel, A. K. 2002. Oscillatory neuronal synchronization in primary visual cortex as a correlate of stimulus selection. *Journal of Neuroscience* 22: 3739–3754.

Gray, C. M. 1999. The temporal correlation hypothesis of visual feature integration: Still alive and well. *Neuron* 24: 31–47.

Gray, C. M., and Singer, W. 1987. Stimulus-specific neuronal oscillations in the cat visual cortex: A cortical functional unit. *Society for Neuroscience Abstracts* 13: 1449.

Gray, C. M., and Singer, W. 1989. Stimulus-specific neuronal oscillations in orientation columns of cat visual cortex. *Proceedings of the National Academy of Sciences of the USA* 86: 1698–1702.

Gray, C. M., König, P., Engel, A. K., and Singer, W. 1989. Oscillatory responses in cat visual cortex exhibit inter-columnar synchronization which reflects global stimulus properties. *Nature* 338: 334–337.

Hebb, D. O. 1949. *The organization of behavior.* New York: Wiley.

Herculano-Houzel, S., Munk, M. H. J., Neuenschwander, S., and Singer, W. 1999. Precisely synchronized oscillatory firing patterns require electroencephalographic activation. *Journal of Neuroscience* 19: 3992–4010.

König, P., Engel, A. K., Löwel, S., and Singer, W. 1993. Squint affects synchronization of oscillatory responses in cat visual cortex. *European Journal of Neuroscience* 5: 501–508.

Keil, A., Müller, M. M., Ray, W. J., Gruber, T., and Elbert, T. 1999. Human gamma band activity and perception of a gestalt. *Journal of Neuroscience* 19: 7152–7161.

Kreiter, A. K., and Singer, W. 1996. Stimulus-dependent synchronization of neuronal responses in the visual cortex of awake macaque monkey. *Journal of Neuroscience* 16: 2381–2396.

Krubitzer, L. 1995. The organization of neocortex in mammals: Are species differences really so different? *Trends in Neuroscience* 18: 408–417.

Krubitzer, L. 1998. Constructing the neocortex: Influences on the pattern of organization in mammals. In M. S. Gazzaniga and J. S. Altman (eds.), *Brain and mind: Evolutionary perspectives* (pp. 18–34). Strasbourg: Human Frontier Science Program Workshop V.

Löwel, S., and Singer, W. 1992. Selection of intrinsic horizontal connections in the visual cortex by correlated neuronal activity. *Science* 255: 209–212.

Miltner, W. H. R., Braun, C., Arnold, M., Witte, H., and Taub, E. 1999. Coherence of gamma-band EEG activity as a basis for associative learning. *Nature* 397: 434–436.

Munk, M. H. J., Roelfsema, P. R., König, P., Engel, A. K., and Singer, W. 1996. Role of reticular activation in the modulation of intracortical synchronization. *Science* 272: 271–274.

Nowak, L. G., Munk, M. H. J., Nelson, J. I., and Bullier, J. A. C. 1995. Structural basis of cortical synchronization, I: Three types of interhemispheric coupling. *Journal of Neurophysiology* 74: 2379–2400.

Philipps, W. A., and Singer, W. 1997. In search of common foundations for cortical computation. *Behavioral Brain Science* 20: 657–722.

Rakiç, P. 1998. Cortical development and evolution. In M. S. Gazzaniga and J. S. Altman (eds.), *Brain and mind: Evolutionary perspectives* (pp. 34–42). Strasbourg: Human Frontier Science Program Workshop V.

Rodriguez, E., George, N., Lachaux, J.-P., Martinerie, J., Renault, B., and Varela, F. J. 1999. Perception's shadow: Long-distance synchronization of human brain activity. *Nature* 397: 430–433.

Roelfsema, P. R., Engel, A. K., König, P., and Singer, W. 1997. Visuomotor integration is associated with zero time-lag synchronization among cortical areas. *Nature* 385: 157–161.

Schmidt, K. E., Goebel, R., Löwel, S., and Singer, W. 1997. The perceptual grouping criterion of colinearity is reflected by anisotropies of connections in the primary visual cortex. *European Journal of Neuroscience* 9: 1083–1089.

Sharma, J., Angelucci, A., and Sur, M. 2000. Induction of visual orientation modules in auditory cortex. *Nature* 404: 841–847.

Singer, W. 1990. The formation of cooperative cell assemblies in the visual cortex. *Journal of Experimental Biology* 155: 177–197.

Singer, W. 1993. Synchronization of cortical activity and its putative role in information processing and learning. *Annual Review of Physiology* 55: 349–375.

Singer, W. 1995. Development and plasticity of cortical processing architectures. *Science* 270: 758–764.

Singer, W. 1999. Neuronal synchrony: A versatile code for the definition of relations? *Neuron* 24: 49–65.

Singer, W., and Gray, C. M. 1995. Visual feature integration and the temporal correlation hypothesis. *Annual Review of Neuroscience* 18: 555–586.

Singer, W., Engel, A. K., Kreiter, A. K., Munk, M. H. J., Neuenschwander, S., and Roelfsema, P. R. 1997. Neuronal assemblies: Necessity, signature and detectability. *Trends in Cognitive Sciences* 1: 252–261.

Steinmetz, P. N., Roy, A., Fitzgerald, P. J., Hsiao, S. S., Johnson, K. O., and Niebur, E. 2000. Attention modulates synchronized neuronal firing in primate somatosensory cortex. *Nature* 404: 187–190.

Stoner, G. R., Albright, T. D., and Ramachandran, V. S. 1990. Transparency and coherence in human motion perception. *Nature* 344: 153–155.

Tallon-Baudry, C., Bertrand, O., Delpuech, C., and Pernier, J. 1997. Oscillatory γ-band (30–70 Hz) activity induced by a visual search task in humans. *Journal of Neuroscience* 17: 722–734.

Tallon-Baudry, C., Bertrand, O., Peronnet, F., and Pernier, J. 1998. Induced γ-band activity during the delay of a visual short-term memory task in humans. *Journal of Neuroscience* 18: 4244–4254.

Tiitinen, H., Sinkkonen, J., Reinikainen, K., Alho, K., Lavikainen, J., and Naatanen, R. 1993. Selective attention enhances the auditory 40-Hz transient response in humans. *Nature* 364: 59–60.

Treisman, M., and Gelade, G. 1980. A feature-integration theory of attention. *Cognitive Psychology* 12: 97–136.

von der Malsburg, C. 1999. The what and why of binding: The modeler's perspective. *Neuron* 24: 95–104.

von der Malsburg, C., and Schneider, W. 1986. A neural cocktail-party processor. *Biological Cybernetics* 54: 29–40.

10 Uniquely Human Cognition Is a Product of Human Culture

Michael Tomasello

We recently completed the Decade of the Brain. During this time, the field of cognitive neuroscience made significant advances in identifying and locating the material bases of many aspects of human behavior and cognition. However, it must not be forgotten that discovering where in the human brain a particular behavioral or cognitive function is localized is only one piece of a large and complex puzzle. As Aristotle explicated long ago, material causes constitute only one of several types of causes that together account for why something is the way that it is.

Following in the tradition of a long line of otherwise diverse thinkers—including Charles Peirce, G. H. Mead, Lev Vygotsky, and Ludwig Wittgenstein—I believe that to fully explain human behavior and cognition we need to go outside of the brain, indeed outside of the skin altogether. My proposal is that there are processes of social interaction and communication occurring between and among human beings that create new behavioral and cognitive functions, in both historical and ontogenetic time. Historically, social interactions enable the creation of cultural artifacts that mediate and amplify human cognition, including such things as linguistic symbols and constructions, Arabic numerals and computers, the Greek alphabet and books, photographs and maps, and on and on, with a somewhat different list for different cultures. Ontogenetically, child–adult social interactions mediated by these symbolic and material artifacts constitute the ecological niche within which normal human cognition, especially in its species-unique aspects, is designed to grow and develop, and so it simply could not come into existence without them (Kruger and Tomasello 1996). Although the structure of the human brain does in some sense enable these historical and ontogenetic interactions to occur, of course, it does not determine these interactions in all of their specificity ahead of time.

Let us be concrete and imagine a mythical wild child growing up completely alone on a desert island—with no language, no mathematical symbols, no pictures or books or other kinds of physical symbols, no adults as teachers, no adults to imitate, no tools or other material artifacts, and so on. What kind of cognition would develop in this child who simply wandered around learning things through its own direct experience of the world, with no access to any cultural tools or symbols? The hypothesis is that it would be much like the cognition of the other great ape species. There would be some uniquely human qualities, for certain, perhaps in such things as making tools and planning action. However, at maturity this individual would have virtually none of the especially flexible and powerful cognitive skills and symbolic representations that make human cognition such a remarkable phenomenon in the natural world. The reason is simply that, to repeat, the species-unique aspects of human cognition develop ontogenetically in the medium of social interactions with other persons and their artifacts, and to deprive it of this medium would be akin to depriving the developing visual system of light. This hypothesis—the cultural origins of uniquely human cognition (see Tomasello 1999a)—is broadly consistent with modern behavioral ecological theories, which stress that uniquely primate and uniquely human cognition evolved first as ways of dealing with complex social problems of cooperation and competition with conspecifics (Humphrey 1976; Byrne and Whiten 1988; Tomasello and Call 1997).

Thus, I believe that to fully account for human cognition—in both its general primate and its species-unique aspects—we need an interacting set of explanations in three distinct time frames: phylogenetic, cultural-historical, and ontogenetic. The hard part is in attempting to specify which aspects of human cognition emanate from which processes in which of these time frames since, in effect, they all work together. Nevertheless, in this chapter I will try to sort out some things in each of these time frames, with a special focus on those aspects of human cognition that are unique to the species.

The Biological Adaptation for Culture Is Uniquely Human

Foundationally, all primates, including humans, perceive the world in the same basic way (Tomasello and Call 1997). All primates share the sensori-motor world of objects in their spatial, temporal, categorical, and quantitative relations, and the social world of behaving conspecifics in their vertical (dominance) and horizontal (affiliative) relationships. And all primate species use their skills and knowledge to formulate creative and insightful strategies when problems arise in either the physical or social domain. Naturally, however, any one species of primate may have additional cognitive skills on top of those shared with other members of the order, and humans are no exception. In the current hypothesis, human beings do indeed possess a species-unique cognitive adaptation—what we might loosely call (without any preconceptions about specific adaptive scenarios) the human adaptation for culture.

The human adaptation for culture most likely occurred after the genus *Homo* split from the genus *Pan* and began its own singular evolutionary trajectory. This hypothesis may seem to contradict a number of recent reports of chimpanzee culture (e.g.,

Whiten et al. 1999), but it does not, if we simply posit that chimpanzee and human cultures each have some characteristics unique to the species. In particular, Tomasello et al. (1993) proposed that what distinguishes human culture from that of chimpanzees and other species is the existence of the "ratchet" effect. The basic idea is that the cultural traditions and artifacts of human beings accumulate modifications over time. Basically none of the most complex human artifacts or social practices—including tool industries, symbolic artifacts, and social institutions—were invented once and for all at a single moment by any one individual or group of individuals. Rather, what happened was that some individual or group of individuals first invented a primitive version of the artifact or practice, and then some later user or users made a modification, an improvement, that others then adopted perhaps without change for many generations, at which point some other individual or group of individuals made another modification, which was then learned and used by others, and so on over historical time. This process of cumulative cultural evolution requires not only creative invention but also, and just as important, faithful social transmission that can work as a ratchet to prevent backward slippage, so that the newly invented artifact or practice may preserve its new and improved form at least somewhat faithfully until a further modification or improvement comes along. The outcome is that human beings are able to pool their cognitive resources in ways that other animal species, whose cultural traditions do not ratchet up in complexity over historical time, are not.

Perhaps surprisingly, for many animal species it is not the creative component but rather the stabilizing ratchet component that is the difficult feat. Thus, many nonhuman primate individuals regularly produce intelligent behavioral innovations and novelties, but then their group mates do not engage in the kinds of social learning that would enable, over time, the cultural ratchet to do its work (Kummer and Goodall 1985). In an effort to explain this difference, Tomasello et al. (1993) distinguished human cultural learning from more widespread forms of social learning and identified three basic types: imitative learning, instructed learning, and collaborative learning. These three types of cultural learning are made possible by a single special form of social cognition, namely, the ability of individual organisms to understand conspecifics as beings like themselves who have intentional and mental lives like themselves. This understanding enables individuals to imagine themselves "in the mental shoes" of some other person, so that they can learn not just from the other, but through the other's perspective. This understanding of others as intentional beings like the self is crucial in human cultural learning because cultural artifacts and social practices—exemplified prototypically by the use of tools and linguistic symbols— invariably point beyond themselves to other outside entities. Tools point to the problems they are designed to solve and linguistic symbols point to the phenomena they are designed to indicate. Therefore, to socially learn the conventional use of a tool or a symbol, children must come to understand why, toward what outside end, the other person is using the tool or symbol; that is to say, the intentional significance of the tool use or symbolic practice—what it is for, what we, the users of this tool or symbol, do with it (Tomasello 1999b).

Processes of cultural learning are especially powerful forms of social learning because they constitute both (1) especially faithful forms of cultural transmission, creating an especially powerful cultural ratchet, and (2) especially powerful forms of social-collaborative creativeness and inventiveness (i.e., processes of sociogenesis in which multiple individuals create something together that no individual could have created on its own). These special powers come directly from the fact that as one human being is learning "through" another, he or she identifies with that other person and his or her intentional and sometimes mental states. Despite some observations suggesting that some nonhuman primates in some situations are capable of understanding con-specifics as intentional agents and of learning from them in ways that resemble some forms of human cultural learning, the overwhelming weight of the empirical evidence suggests that in their natural habitats only human beings understand conspecifics as intentional agents like the self and so only human beings engage in cultural learning (see Tomasello 1996; Tomasello and Call 1997).

The most plausible conclusion is thus that the social learning skills of chimpanzees are sufficient to create and maintain their species-typical cultural activities, but they are not sufficient to create and maintain humanlike cultural activities displaying the ratchet effect and cumulative cultural evolution (Boesch and Tomasello 1998). It is worth noting in this regard that in their natural habitat, chimpanzees' sister species, bonobos (*Pan paniscus*), have not so far been observed to show anything resembling the "cultural" traditions of chimpanzees. Applying the basic inferential techniques of comparative biology, this fact suggests that the common ancestor of humans and these two sister species did not have humanlike cultural learning skills either. Indeed, examining the record of human evolution, it seems most likely that early humans did not have them for most of their history. Most likely, skills of cultural learning and the resulting cumulative cultural evolution began only with modern humans, as they first evolved somewhere in Africa some 200,000 years ago, and this adaptational event may even explain why modern humans outcompeted other hominoids as they migrated all over the globe (Klein 1998; Foley and Lahr 1997).

Uniquely Human Cognition Is a Historical Phenomenon

In general, then, human cultural traditions may be most readily distinguished from chimpanzee cultural traditions—as well as the few other instances of culture observed in other primate species—by the fact that they accumulate modifications over time; that is to say, they have cultural "histories." They accumulate modifications and have histories because the cultural learning processes that support them are especially powerful. They are especially powerful because they are supported by the uniquely human cognitive adaptation for understanding others as intentional beings like the self, which creates forms of cultural learning that act as a ratchet by faithfully preserving newly innovated strategies in the social group until there is another innovation to replace them.

These historical processes may be clearly seen in two of the most fundamental human cognitive domains: language and mathematics. First, it is clear that each of the thousands of languages extant has its own inventory of linguistic symbols and constructions that allow its users to symbolically share experiences with one another. This inventory of symbols and constructions is grounded in universal structures of human cognition and communication, and the mechanics of the vocal-auditory apparatus. Thus all languages share some features, but particular languages each have their own cultural histories. These come from differences among the various peoples of the world in the kinds of things they consider it important to talk about and the ways they think it useful to talk about them, along with various historical accidents, of course. The crucial point for current purposes is that all of the symbols and constructions of a given language are not invented at once, and once invented, they often do not stay the same for very long. Rather, linguistic symbols and constructions evolve and change and accumulate modifications over historical time as human beings use them with one another and adapt them to changing circumstances.

The most important dimension of the historical process in the current context is grammaticization or syntacticization, which involves the congealing of loose and redundantly organized discourse structures into tight and less redundantly organized syntactic constructions (see Traugott and Heine 1991a,b and Hopper and Traugott 1993, for some recent research). For example, (1) loose discourse sequences such as "He pulled the door and it opened" may become syntacticized into "He pulled the door open" (a resultative construction). (2) Loose discourse sequences such as "My boyfriend . . . He plays piano . . . He plays in a band" may become "My boyfriend plays piano in a band." (3) A sequence such as "My boyfriend . . . He rides horses . . . He bets on them" may become "My boyfriend, who rides horses, bets on them." (4) Complex sentences may also derive from discourse sequences of initially separate utterances, as in "I want it . . . I buy it" evolving into "I want to buy it." In the process, free-standing, contentful words often turn into grammatical morphemes (e.g., auxiliaries, prepositions, tense markers, or case markers), as a kind of "glue" that holds the new construction together. Since children now learn these constructions as wholes, we can clearly see the ratchet effect working in language history.

Systematic investigation into processes of grammaticization and syntacticization is still in its infancy, and indeed the suggestion that languages may have evolved from structurally simpler to structurally more complex forms through processes of grammaticization and syntacticization is somewhat novel in this context; linguists most often think of these processes as sources of change only. Nevertheless, grammaticization and syntacticization are able to effect serious changes in linguistic structure in relatively short periods of time—for example, the main diversification of the Romance languages took place during some hundreds of years—and thus there is no reason why they could not also work to make a simpler language more complex syntactically in some thousands of years. Exactly how grammaticization and syntacticization happen in the concrete interactions of individual human beings and groups of human beings, and how these processes might relate to the other processes of sociogenesis through which human social interaction ratchets up in complexity cultural artifacts, is a question for future linguistic research.

The case of the other intellectual pillar of Western civilization, mathematics, is interestingly different from the case of language. Like language, mathematics clearly rests on universally human ways of experiencing the world (many of which are shared with other primates) and also on some processes of cultural creation and sociogenesis. However, in this case the divergences among cultures is much greater than in the case of spoken languages. All cultures have complex forms of linguistic communication, with variations in complexity that are basically negligible, whereas some cultures have highly complex systems of mathematics (practiced by only some of their members) compared with other cultures that have fairly simple systems of numbers and counting (Saxe 1981). This great variation means that no one has proposed that the structure of modern complex mathematics is an innate module, as they have in the case of language.

In general, the reasons for the great cultural differences in mathematical practices are not difficult to discern. First, different cultures and persons have different needs for mathematics. Most cultures and persons have the need to keep track of goods, for which a few number words expressed in natural language will suffice. When a culture or a person needs to count objects or measure things more precisely—for example, in complex building projects, trading practices, or the like—the need for a more complex mathematics arises. Modern science as an enterprise practiced by only some people in some cultures presents a whole host of new problems that require complex mathematical techniques for their solution. However, complex mathematics as we know it today can only be accomplished through the use of certain forms of graphic symbols. In particular, the Arabic system of enumeration is much superior to older Western systems (e.g., Roman numerals) for complex mathematics, and the use of Arabic numerals, including zero and the place system for indicating different-sized units, opened up for Western scientists and other persons whole new vistas of mathematical operations (Danzig 1954).

The history of mathematics is an area of study in which detailed examination has revealed all kinds of complex ways in which individuals and groups of individuals take what is passed on to them by previous generations and then make modifications as needed to deal with new practical and scientific problems more efficiently. Some historians of mathematics have detailed some of the processes by which specific mathematical symbols and techniques were invented, used, and modified (e.g., Danzig 1954; Eves 1961; Damerow 1998). As just one well-known example, Descartes invented the Cartesian coordinate system in which he combined in a creative way some of the spatially based techniques used in geometry with some of the more arithmetically based techniques in other areas of the mathematics of his time, with the infinitesimal calculus being a variation on this theme. The adoption of this technique by other scientists and mathematicians ratcheted up the mathematical universe almost immediately and thereby changed Western mathematics forever. And so, in general, the sociogenesis of modern Western mathematics, as practiced by only a minority of the people in these cultures, may be seen as a function of both the mathematical needs of the particular people involved and the cultural resources they have available to them.

In the case of both language and mathematics, then, the structure of the domain as it now exists has a cultural history (actually, many different cultural histories), and there are processes of sociogenesis that historical linguists and historians of mathematics have the opportunity to study. The differences in the two cases are instructive. Although complexity takes many different forms in modern languages, complex language is a universal among all the peoples of the world. This is either because the original invention of many of the spoken symbols that makes language possible took place before modern humans diverged into different populations, or else the ability to create spoken symbols comes so naturally to humans that the different groups all invented them in similar though not identical ways after they diverged. Complex mathematics is not universal among different cultures, or even among people in the cultures that have it. This is presumably because the cultural needs for complex mathematics and/or the invention of the required cultural resources came only after modern people had begun living in different populations, and apparently these needs and/or resources are not universally present for all peoples of the world today. And so one of the central facts about language that has led some scholars to hypothesize that some modern linguistic structures are innate—the fact that they are species-unique and species-universal, whereas many mathematical and other cognitive skills are not (e.g., Pinker 1994)—may simply be the result of the vagaries of human cultural history in the sense that, for whatever reason, skills of linguistic communication evolved before modern humans diverged into separate populations, whereas those of mathematics did not.

The Ontogeny of Uniquely Human Cognition

The place where human biology meets human cultural history is, of course, human ontogeny. Indeed, there is a specific moment in human ontogeny, at about 9 months of age, where the human adaptation for culture first shows itself. From that point on, as the child interacts with his or her physical and social worlds through the mediating lens of the cultural artifacts and symbols inherited historically, new forms of cognition—with no other precedents in the natural world—begin to emerge.

Six-month-old infants interact dyadically with objects, grasping and manipulating them, and interact dyadically with other people, expressing emotions back-and-forth in a turn-taking sequence. However, at around 9 to 12 months of age, infants begin to engage in interactions that are triadic in the sense that they involve the referential triangle of child, adult, and some outside entity with which they share attention. Thus, infants at this age begin to flexibly and reliably look where adults are looking (gaze following), use adults as social reference points (social referencing), and act on objects in the way adults are acting on them (imitative learning); in short, to "tune in" to the attention and behavior of adults toward outside entities. At this same age, infants also begin to use communicative gestures to direct adult attention and behavior to outside entities in which they are interested; in short, to get the adult to "tune in" to them. This revolution in the way infants relate to their worlds begins when infants understand other persons as intentional agents like the self who have a perspective on the world that can be followed, directed, and shared (Tomasello 1995; Carpenter et al. 1998b).

This social-cognitive revolution at 1 year of age sets the stage for infants' second year of life, in which they begin to imitatively learn the use of all kinds of tools, artifacts, and symbols. For example, in a study by Meltzoff (1988), 14-month-old children observed an adult bend at the waist and touch her head to a panel, thus turning on a light. The infants followed suit. They engaged in this somewhat unusual and awkward behavior even though it would have been easier and more natural for them to simply push the panel with their hand. One interpretation of this behavior is that the infants understood that (1) the adult had the goal of illuminating the light and then chose one means for doing so from among other possible means; and (2) if they had the same goal, they could choose the same means. Cultural learning of this type thus relies fundamentally on infants' tendency to identify with adults and on their ability to distinguish in the actions of others the underlying goal and the different means that might be used to achieve it. This interpretation is supported by a later finding by Meltzoff (1995) that 18-month-old children also imitatively learn actions that adults intend to perform, even if the adults are unsuccessful in doing so. Similarly, Carpenter et al. (1998b) found that 16-month-old infants will imitatively learn from a complex behavioral sequence only those behaviors that appear intentional, ignoring those that appear accidental. Young children do not just mimic the limb movements of other persons, they attempt to reproduce other persons' intended actions in the world.

Although it is not obvious at first glance, something like this same imitative learning process must happen if children are to learn the symbolic conventions of their native language. Although it is often assumed that young children acquire language when adults stop what they are doing, hold up objects, and name these objects for them, this is empirically not the case. Linguistics lessons such as these are characteristic of only some parents in some cultures and for only some kinds of words (i.e., no one names for children acts of "giving" or prepositional relationships such as "on" or "for"). In general, for the greatest part of their language, children must find a way to learn a new word in the continuing flow of social interaction, sometimes from speech not even addressed to them (Brown 2001). In some recent experiments, something of this process has been captured. For example, in the context of a finding game, an adult announced her intentions to "find the toma" and then searched in a row of buckets that all contained novel objects. Sometimes she found it in the first bucket searched, smiling and handing the object to the child. Sometimes, however, she had to search longer, rejecting unwanted objects by scowling at them and replacing them in their buckets until she found the one she wanted (again indicated by a smile and the termination of the search). The children learned the new word for the object the adult intended to find regardless of whether or how many objects were rejected during the search process (Tomasello and Barton 1994; see Tomasello 2001, for a review of other similar studies). Indeed, a strong argument can be made that children can understand a symbolic convention in the first place only if they understand their communicative partner as an intentional agent with whom one may share attention, since a linguistic symbol is nothing other than a marker for an intersubjectively shared understanding of a situation (Tomasello 1998, 2001).

What we are witnessing here, in both the instrumental and linguistic domains, is nothing other than the ontogenetic emergence of the human adaptation for culture as it meets the historically evolved material and symbolic artifacts of the culture. As a point of comparison, we may invoke the case of children with autism, who do not seem to understand other persons as intentional agents (or else they do so only to an imperfect degree). Autistic children are thus very poor at the imitative learning of intentional actions in general; only half of them ever learn any language at all, and those who do learn some language are poor in word-learning situations such as those just described. Because of a biological deficit in the human adaptation for culture, autistic children are not able, as are typically developing children, to tune into the accumulated tools and symbols of their culture, and so they remain to some degree acultural beings (Hobson 1993; Baron-Cohen 1993).

Tomasello (1999a) hypothesized that as a result of engaging for many years in cultural learning processes in which they attempted to see the world and to learn about it from the perspective of other persons—especially in the case of language—human children develop a species-unique form of cognitive representation. The basic idea is this: The symbolic artifacts that comprise English, Turkish, or any language, are the result of a particular group of people inventing and modifying over historical time ways for sharing and manipulating attention. When today's child learns the conventional use of these well-traveled symbols, what the child is learning are the ways his or her forebears in the culture found it useful to share and manipulate the attention of others in the past. And because the peoples of a culture, as they move through historical time, evolve many and varied purposes for manipulating the attention of one another, and because they need to do this in many different types of discourse situations, today's child is faced with a whole panoply of different linguistic symbols and constructions that embody many different attentional construals of any given situation. As just a sampling, languages embody attentional construals based on such things as (see Langacker 1987 for more specifics):

- granularity-specificity (thing, furniture, chair, desk chair)
- perspective (chase-flee, buy-sell, come-go, borrow-lend)
- function (father, lawyer, man, American) (coast, shore, beach).

There are many more specific perspectives that arise in grammatical combinations of various sorts ("She smashed the vase" versus "The vase was smashed"). It is at about 18 months of age that children first begin to predicate multiple things about objects to which they and an adult are jointly attending, for example, by saying that this ball is either wet or big or mine, which is all about one and the same object (Tomasello 1988, 1995).

Consequently, as the young child internalizes a linguistic symbol, as she learns the human perspective embodied in that symbol, she cognitively represents not just the perceptual or motoric aspects of a situation, but cognitively represents one way, among other ways of which the child is also aware, that the current situation may be attentionally construed by us, the users of the symbol. The way that human beings use linguistic symbols thus creates a clear break with straightforward perceptual or sensorimotor cognitive representations that are characteristic of human infants and other animal species. It is true that a prelinguistic child, or a nonhuman primate, may construe situations in more than one way. One time a conspecific is a friend and the next time an enemy; one time a tree is for climbing to avoid predators and the next time it is for making a nest in. In these different interactions with the same entity, the individual is deploying its attention differentially, depending on its goal at that moment. However, shifting attention sequentially in this manner as a function of a goal is not the same thing as knowing simultaneously a number of different ways in which something might be construed. An individual user of language looks at a tree and before drawing the attention of an interlocutor to that tree, must decide, based on an assessment of the listener's current knowledge and expectations, whether to use "That tree over there, It, The oak, That hundred-year-old oak, The tree, The bagswing tree, That thing in the front yard, The ornament, The embarrassment", or any number of other expressions. The individual must decide if the tree "is in/is standing in/is growing in/was placed in/is flourishing" in the front yard.

Decisions such as these are not made on the basis of the speaker's direct goal with respect to the object or activity involved, but on the basis of the speaker's goal with respect to the listener's interest and attention to that object or activity. This means that the speaker knows that the listener shares with him or her these same choices for construal, again, all of which are available simultaneously. Indeed, the fact is that while she or he is speaking, a speaker is constantly monitoring the listener's attentional status (and vice versa), which means that both participants in a conversation are always aware that there are at least two actual perspectives on a situation, as well as many more perspectives that can be symbolized in currently unused symbols and constructions. Human linguistic symbols are thus both intersubjective—shared devices for sharing and manipulating attention—and perspectival in the sense that any one symbol embodies one way, out of many other simultaneously available ways, that a situation may be construed (Tomasello 1999a).

Some of the effects of operating with symbols of this type are obvious, in terms of flexibility and relative freedom from perception. However, some are more far-reaching and quite unexpected, I think, in the sense that they give children truly new ways of conceptualizing things, such as treating objects as actions ("He porched the newspaper"), actions as objects ("Skiing is fun"), and all kinds of metaphorical construals of things ("Love is a journey") (Lakoff and Johnson 1980; Lakoff 1987; Johnson 1987). These new ways of conceptualizing and thinking result from the accumulated effects of engaging in linguistic communication with other persons for some years during early cognitive development. More extended bouts of communication with other persons also create opportunities for explicitly exploring and comparing differing verbally expressed perspectives on situations. Perhaps of special importance are interactions in which the communicative partner provides a verbally expressed perspective on the child's previous verbally expressed perspective, since in this case the internalization of the other's perspective helps to create children's ability to self-regulate, self-monitor, and reflect on their own cognition (Vygotsky 1978).

The point is not just that linguistic symbols provide handy tags for human concepts or even that they influence or determine the shape of those concepts, although they do both of these things. The point is that the intersubjectivity of human linguistic symbols—and their perspectival nature as one offshoot of this intersubjectivity—means that linguistic symbols do not represent the world directly, in the manner of perceptual or sensorimotor representations, but rather are used by people to induce others to construe certain perceptual and conceptual situations—to attend to them—in one way rather than in another. This breaks these symbols away from the sensorimotor world of objects in space and puts them instead into the realm of the human ability to view the world in whatever way is useful for the communicative purpose at hand. The most important point in the current context is that as children participate in these communicative exchanges, they internalize, in something like the way

Vygotsky (1978) envisioned the process, the perspectives of other persons as embodied in linguistic symbols and constructions. A number of investigators have argued and presented evidence that this process is so powerful that children learning different languages end up with somewhat different conceptions of basic aspects of their physical and social worlds (see Gumperz and Levinson 1996, for some representative examples).

Conclusion

Modern cognitive science is moving more and more in the direction of biological explanations for human cognition, both in the sense of evolutionary speculations (e.g., Pinker 1997) and neurophysiological reductions (Gazzaniga 1999). These are both important lines of research, of course, and any full account of human cognition will have to include them both. The only problem is that these approaches are often myopic in their reductionism, to the point that they basically ignore altogether the social-interactional and cultural dimensions of human cognition, which arguably constitute its most distinctive features. The fact that human biology enables the kinds of social, cultural, and linguistic interactions that are characteristic of human beings does not mean that it determines them in all of their specificities ahead of time. Indeed, the existence of the many varieties of human behavior evident in the different cultures of the world demonstrates conclusively that it does not. It is obvious to anyone who cares to look at them that these social, cultural, and linguistic interactions—especially because they result in important cultural artifacts—play a crucial, indeed a constitutive, role in the structure of uniquely human processes of cognition.

My account is thus that there was a single cognitive adaptation, originating with modern humans, that accounts for most, if not all, of the many differences in human and nonhuman primate cognition. This single adaptation made possible an evolutionarily new set of processes of sociogenesis, operating in historical and ontogenetic time, that have done much of the actual work in creating the symbolic representations that give human cognition much of its awesome power. In effect, this adaptation enabled the development of new forms of social interaction and cognitive representation, which over time transformed such basic primate competencies as intentional communication, social dominance, social exchange, and cognitive exploration, into the uniquely human cultural institutions of language, government, money, and science—without any additional genetic events. Let me be clear. There are certainly human cognitive functions in which historical and ontogenetic processes play only a minor role; for example, the basic processes of perceptual categorization, quantification, spatial understanding, object permanence, social perception, social learning, and so on. Social-interactive processes do not create these and similar functions; they are part of primate cognition in general (Tomasello and Call 1997).

However, such things as linguistic symbols and cultural institutions are socially constituted and culturally variable, and so human forms of social-interactive processes, working in combination with basic processes of primate cognition, must have played a crucial role in their creation and historical evolution.

Overall, the basic problem with all exclusively biological approaches to human cognition, especially when they address its uniquely human aspects, is that they attempt to skip from the first page of the story, genetics, to the last page of the story, current human cognition, without going through any of the intervening pages. In my opinion, any adequate theory of human cognition must provide some reasonable account of the processes of sociogenesis in historical and ontogenetic time that intervened between the human genotype and the human phenotype. This is quite simply because it is these processes, and not any biological events directly, that have created uniquely human material and symbolic artifacts, which in turn have created an evolutionarily unique cultural niche for human cognitive ontogeny, which in turn has created an evolutionarily new form of cognition that relies on intersubjective and perspectival cognitive representations.

Note

This chapter was written in the fall of 2000.

References

Baron-Cohen, S. 1993. From attention-goal psychology to belief-desire psychology. In S. Baron-Cohen, H. Tager-Flusberg, and D. Cohen (eds.), *Understanding other minds* (pp. 59–82). Oxford: Oxford University Press.

Boesch, C., and Tomasello, M. 1998. Chimpanzee and human culture. *Current Anthropology* 39: 591–614.

Brown, P. 2001. Learning to talk about motion UP and DOWN in Tzeltal. In M. Bowerman and S. C. Levinson (eds.), *Language acquisition and conceptual development* (pp. 512–543). Cambridge: Cambridge University Press.

Byrne, R. W., and Whiten, A. 1988. *Machiavellian intelligence. Social expertise and the evolution of intellect in monkeys, apes, and humans.* New York: Oxford University Press.

Carpenter, M., Nagell, K., and Tomasello, M. 1998a. Social cognition, joint attention, and communicative competence from 9 to 15 months of age. *Monographs of the Society for Research in Child Development*, vol. 63.

Carpenter, M., Akhtar, N., and Tomasello, M. 1998b. Fourteen-through 18 month-old infants differentially imitate intentional and accidental actions. *Infant Behavior and Development* 21(2): 315–330.

Damerow, P. 1998. Prehistory and cognitive development. In J. Langer and M. Killen (eds.), *Piaget, evolution, and development* (pp. 247–269). Mahwah, N.J.: Erlbaum.

Danzig, T. 1954. *Number: The language of science*. New York: Free Press.

Eves, H. 1961. *An introduction to the history of mathematics*. New York: Holt, Rinehart, & Winston.

Foley, R., and Lahr, M. 1997. Mode 3 technologies and the evolution of modern humans. *Cambridge Archeological Journal* 7: 3–36.

Gazzaniga, M. 1999. *The new cognitive neurosciences*. Cambridge, Mass.: MIT Press.

Gumperz, J., and Levinson, S. 1996. *Rethinking linguistic relativity*. Cambridge: Cambridge University Press.

Hobson, P. 1993. *Autism and the development of mind*. Hillsdale, N.J.: Erlbaum.

Hopper, P., and Traugott, E. 1993. *Grammaticalization*. Cambridge: Cambridge University Press.

Humphrey, N. K. 1976. The social function of intellect. In P. Bateson and R. A. Hinde (eds.), *Growing points in ethology* (pp. 303–321). Cambridge: Cambridge University Press.

Johnson, M. 1987. *The body in the mind*. Chicago: University of Chicago Press.

Klein, R. 1998. *The human career: Human biological and cultural origins* (2nd ed.). Chicago: University of Chicago Press.

Kruger, A., and Tomasello, M. 1996. Cultural learning and learning culture. In D. Olson (ed.), *Handbook of education and human development: New models of teaching, learning, and schooling* (pp. 169–187). Oxford: Blackwell.

Kummer, H., and Goodall, J. 1985. Conditions of innovative behaviour in primates. *Philosophical Transactions of the Royal Society of London* B308: 203–214.

Lakoff, G. 1987. *Women, fire, and other dangerous things: What categories reveal about the mind*. Chicago: University of Chicago Press.

Lakoff, G., and Johnson, M. 1980. *Metaphors we live by*. Chicago: University of Chicago Press.

Langacker, R. 1987. *Foundations of cognitive grammar*, Vol. 1. Stanford, Calif.: Stanford University Press.

Meltzoff, A. 1988. Infant imitation after a one-week delay: Long-term memory for novel acts and multiple stimuli. *Developmental Psychology* 24: 470–476.

Meltzoff, A. 1995. Understanding the intentions of others: Re-enactment of intended acts by 18-month-old children. *Developmental Psychology* 31: 838–850.

Pinker, S. 1994. *The language instinct: How the mind creates language*. New York: Morrow.

Pinker, S. 1997. *How the mind works*. London: Penguin.

Saxe, G. 1981. Body parts as numerals: A developmental analysis of numeration among a village population in Papua New Guinea. *Child Development* 52: 306–316.

Tomasello, M. 1988. The role of joint attentional process in early language development. *Language Sciences* 10: 69–88.

Tomasello, M. 1995. Joint attention as social cognition. In C. Moore, and P. Dunham (eds.), *Joint attention: Its origins and role in development* (pp. 103–130). Hillsdale, N.J.: Erlbaum.

Tomasello, M. 1996. Do apes ape? In B. G. Galef, Jr. and C. M. Heyes (eds.), *Social learning in animals: The roots of culture* (pp. 319–346). New York: Academic Press.

Tomasello, M. 1998. Reference: Intending that others jointly attend. *Pragmatics and Cognition*, 6: 219–234.

Tomasello, M. 1999a. *The cultural origins of human cognition*. Cambridge, Mass.: Harvard University Press.

Tomasello, M. 1999b. The cultural ecology of young children's interactions with objects and artifacts. In E. Winograd, R. Fivush, and W. Hirst (eds.), *Ecological approaches to cognition: Essays in honor of Ulric Neisser* (pp. 141–170). Mahway, N.J.: Erlbaum.

Tomasello, M. 2001. Perceiving intentions and learning words in the second year of life. In M. Bowerman and S. C. Levinson (eds.), *Language acquisition and conceptual development* (pp. 132–158). Cambridge: Cambridge University Press.

Tomasello, M., and Barton, M. 1994. Learning words in non-ostensive contexts. *Developmental Psychology* 30: 639–650.

Tomasello, M., and Call, J. 1997. *Primate cognition*. Oxford: Oxford University Press.

Tomasello, M., Kruger, A. C., and Ratner, H. H. 1993. Cultural learning. *Behavioral and Brain Sciences* 16: 495–552.

Traugott, E., and Heine, B. 1991a. *Approaches to grammaticalization*. Vol. 1. Amsterdam: John Benjamins.

Traugott, E., and Heine, B. 1991b. *Approaches to grammaticalization*. Vol. 2. Amsterdam: John Benjamins.

Vygotsky, L. 1978. *Mind in society: The development of higher psychological processes*. Cambridge, Mass.: Harvard University Press.

Whiten, A., Goodall, J., McGrew, W., Nishida, T., Reynolds, V., Sugiyama, Y., Tutin, C., Wrangham, R., and Boesch, C. 1999. Cultures in chimpanzees. *Nature* 399: 682–685.

11 Moral Ingredients: How We Evolved the Capacity to Do the Right Thing

Marc D. Hauser

The Moral Menu

In *David Copperfield*, a poignant novel about human nature, Charles Dickens takes the reader on an adventure filled with rich moral dilemmas. Even before David himself has any sense of who he is or what he will become, we are told that his great-aunt, Betsey Trotwood, is disappointed in his gender—she had hoped for a baby girl—and that he is fatherless. He does, however, have a loving mother and an exceptionally caring servant in Miss Peggotty. David's mother eventually remarries, and this puts David, now a young boy, in a terrible bind. David loves his mother, but she has married a man—Mr. Murdstone—of considerable greed and not a shred of compassion. Although David tries to do the right thing, his stepfather is like a lion taking over a pride and attempting to do away with all the cubs. He has no use for David and at the outset wants to remove him from the house if not life in general. In a passage that is certainly a high point for the reader, David bites his stepfather's hand while in the midst of receiving a completely unwarranted flogging. This violation of respectful conduct enrages his stepfather, who proceeds to send him off to boarding school. Unfortunately for poor David, this is only the beginning of his travails because each developmental turn is met with yet another moral twist and challenge. Such is life.

To understand the moral weight of *David Copperfield*, or our moral sense more generally, I argue that we need to perform two conceptual tasks. First, we must follow the lead of Immanuel Kant (1781/1929, 1797/1965), who argued that morality should be examined like a chemist looking at a crystal; take the whole and break it down into its constituent parts. Thus, in evaluating the moral thread in *David Copperfield*, we might ask why it seems wrong for Betsey Trotwood to be disappointed in David's gender. Doesn't she have the right to a preference? Why do we feel that David is justified in biting his stepfather, Mr. Murdstone? Why couldn't he inhibit this attack? What makes Murdstone greedy and why do we, as consumers of such literature, think that it is wrong, a character flaw? Why doesn't David's mother have the right to counter Murdstone's desire to send David off to boarding school? Why should her loyalty lie with him rather than with her son? In answering each of these questions, we are not only seeking a theoretically motivated answer, but an answer that begins to pinpoint how our brains and minds achieve such answers and thus motivate our moral actions. Second, we must evaluate morality as Darwin (1871, 1872) did in the late 1800s, and would do today if he were to have access to the rich comparative work on animals, from brain to mind to behavior. We must look at the design features of our moral sense and explore the extent to which such features are shared with other animals.

Kant's reductionist approach to morality aims to uncover the ingredients that constitute a moral system. Darwin's evolutionary approach aims to uncover why moral systems are built the way they are, and in what way they are adaptive. As I have argued elsewhere (Hauser 2000), morality, like language, is guided by a suite of universal mechanisms that limit the range of variation, and thus the range of possible moral systems. In the same way that Chomsky (1957, 1986), Jackendoff (1994), Pinker (1994), and others have argued that language is mediated by innate mechanisms that limit linguistic variation, I believe that the same argument can be made for morality. I am certainly not alone in this endeavor (Alexander 1987; Damasio 1994; de Waal 1996; Ridley 1996; Sober and Wilson 1998; H. Q. Wilson 1993; Wright 1994). To illustrate the fertility of this approach, I will explore in detail one small, albeit important ingredient of our moral sense: the capacity to inhibit actions in the face of several possible alternatives. In the second section, I begin by making a distinction between inhibitory mechanisms that are responsible for emotionally salient as opposed to conceptually challenging decisions. I develop these ideas by reviewing some of the relevant literature from child development, cognitive neuroscience, and animal behavior. In the third section, I turn to cases of animal behavior that, although limited relative to human actions, provide the appropriate flavor with respect to our concerns about morality. In particular, I review experiments that explore whether animals have a sense of "doing the right thing," of helping those in need or punishing those who attempt to cheat.

Decisions, Perseveration, and the Brain

When David Copperfield brings James Steerforth—an old boarding school chum—back to visit his childhood friends, he is confronted with yet another moral dilemma. Steerforth is a suave, confident, and handsome man and is clearly attracted to David's childhood playmate Em'ly, who is poor and engaged to be married to Ham (Peggotty's nephew), a wonderful man and family friend. David sees the writing on the wall, but is emotionally torn between warning Em'ly and Ham and keeping his mouth shut to honor his friendship. This moral dilemma is a classic in revealing the signature design of all moral situations: a pull between emotionally conflictual decisions or outcomes.

Ultimately, one has to choose. David does. He keeps quiet, Steerforth sweeps Em'ly off her feet, they elope, Ham is emotionally leveled, and David feels absolutely awful.

Such decision-making is complicated. However, if we filter some of the layers of complexity, we can explore a simpler problem: How do individuals develop the capacity to inhibit actions toward emotionally or motivationally prepotent goals? An answer comes from comparative studies of human and nonhuman primate development at both the behavioral and neurobiological level.

When Passion Gets in the Way

Ever since Piaget (1952), developmental psychologists have known that young children often perseverate with a behavioral response, doing the same thing over and over again. For example, if a young infant watches as a novel toy is concealed in location A and not B, the infant will retrieve the object at A, and will do so as long as it is concealed there. If the infant now watches as the object is concealed in location B, it will continue to search for the object at A, not B. This perseverative error is robust (i.e., repeatable over dozens of trials with the same subject) up until the age of approximately 7–9 months; older infants generally search for the object in the correct location following the switch. Developmentalists have argued that this error occurs because young infants have difficulty inhibiting a prepotent response bias, one that leads them to search at the location that was previously reinforced.

To investigate the ontogeny of this inhibitory mechanism, Diamond and colleagues (Diamond 1988, 1990; Diamond and Goldman-Rakic 1989; Diamond et al. 1989) conducted an elegant series of experiments with human infants as well as rhesus monkey infants and adults. The basic task involved presenting subjects with a novel object (a toy for infants, food for the monkeys) placed inside a transparent box with one open side. For most subjects, the first response or so involved reaching straight ahead for the object. Consequently, the subjects reached straight into a Plexiglas wall, failing to obtain the desired object. Three groups of subjects continued with this straight-ahead reaching response: infants under the age of 7–9 months, 2–4-month-old rhesus infants, and adult rhesus monkeys with lesions in the dorsolateral prefrontal cortex (figure 11.1). In contrast, the older human infants and the adult rhesus monkeys had no problem with this task, reaching into the open side of the box to retrieve the target object. Based on these results, Diamond and her colleagues drew two relevant conclusions: First, the capacity to inhibit prepotent emotional response biases takes time to mature in both humans and rhesus monkeys. Second, an important locus of control for such inhibitory processes is the dorsolateral prefrontal cortex.

Diamond's experiments are important, for they show the power of the comparative method, of looking at developmental change, and of uncovering the neural substrates guiding behavior. However, the comparative method's power comes from a much broader taxonomic sweep, contrasting both closely and distantly related species. With

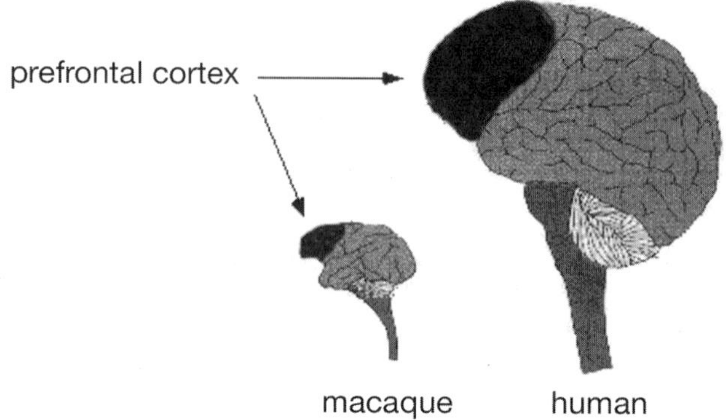

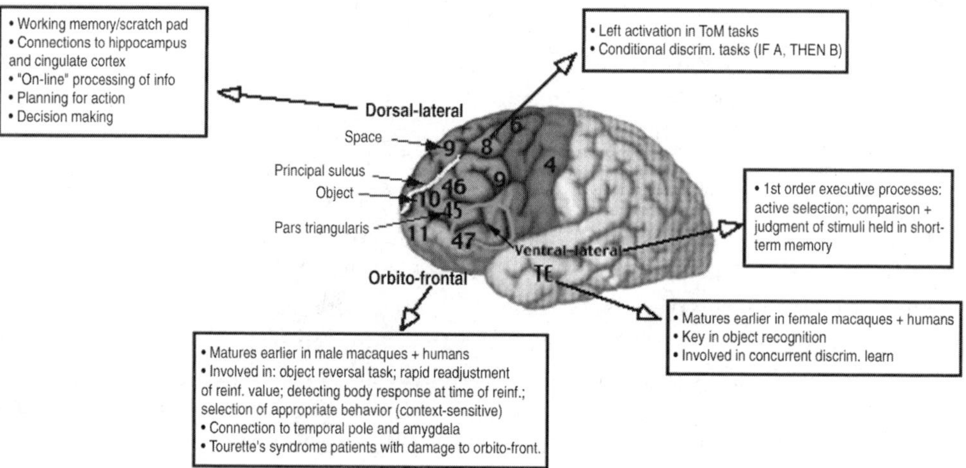

Figure 11.1
The top panel illustrates the overall difference in the proportion of brain dedicated to the prefrontal cortex in macaques and humans. The lower panel shows some of the key anatomical
landmarks in the prefrontal cortex and some of the associated functions.

this goal in mind, my students and I (Santos et al. 1999) set out to use Diamond's object retrieval task with adult cotton-top tamarins, a New World monkey species (*Saguinus oedipus*). To give a sense of time depth, whereas the evolutionary distance between rhesus monkeys and humans is approximately 20–30 million years, we shared a common ancestor with tamarins approximately 40–60 million years ago. Thus, although tamarins are also primates, they begin to push the comparative problem farther back in time.

We divided our tamarin colony into two groups. One group was tested on Diamond's original task involving a Plexiglas box with one open side (figure 11.2). We fully expected these subjects to pass the test, given that they were adults with intact, mature dorsolateral prefrontal cortices. To our surprise, these subjects never reached criterion, continuing to reach straight ahead for the food reward and consequently banging into the Plexiglas wall. The second group started out with an opaque box that matched the dimensions of the transparent one and consisted of only one open side. The subjects watched as the opaque box was lowered over a piece of food and then are moved forward, providing the subjects with an opportunity to reach the box. In this condition, the subjects readily reached around to the open side and retrieved the food reward. In the second condition, we presented the subjects with a transparent box with red gridlines that were designed to help demarcate the edges and sides of the box. Once again, the subjects reached around to the open side and retrieved the food. In the final condition, we presented the subjects with the original transparent box and here too they reached around to the open side and retrieved the food.

Although there were no differences in age or experimental history in the subjects in these two groups, the tamarins in the first group performed like young human infants, rhesus infants, and adult rhesus with lesions to the dorsolateral prefrontal cortex. In contrast, the tamarins in the second group performed like older human infants and intact adult rhesus monkeys. The results from the second group are particularly striking. If tamarins have a general problem inhibiting prepotent emotions or motives, then when the subjects were transferred from the opaque to the transparent box condition, they should have reached straight ahead and perseverated with this response bias. The tamarins did not, however, respond in this way. Comparable results have been reported by Roberts and her colleagues (Roberts et al. 1998; Roberts and Wallis 2000) for the closely related common marmoset.

Our replication of Diamond's object retrieval task, together with the work of Roberts and colleagues on the common marmoset, allows us to make at least four points. First, it is important to avoid drawing the conclusion that if one primate species does something, all primate species will do the same. Diamond doesn't draw this conclusion, but many neuroscientists are prone to talking about results in "the monkey," a species that simply does not exist (Hauser 1999; Preuss 1995). Second, although Diamond is certainly on solid ground in showing an important maturational shift in the

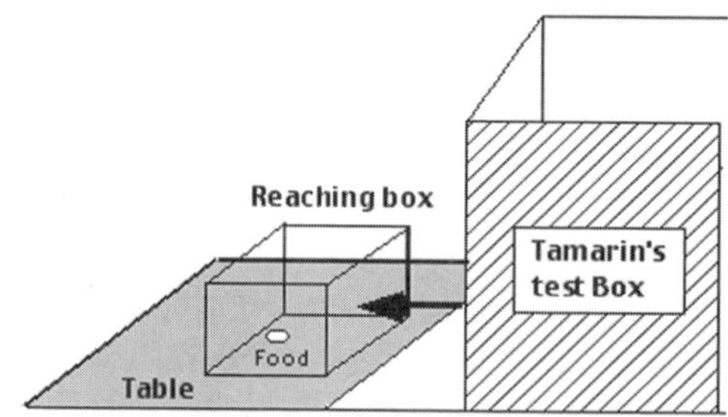

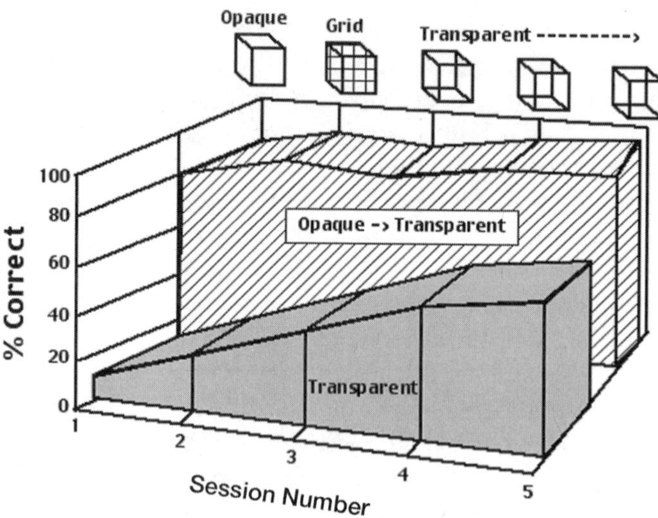

Figure 11.2

Performance of cotton-top tamarins on Diamond's object retrieval task. There were two groups. One group was tested on a transparent box with one open side. The second group was first tested on an opaque box with one open side and then transferred to a transparent box with one open side.

functioning of the dorsolateral prefrontal cortex, this developmental change need not account for all perseverative errors. The tamarins that we tested were adults. Thus, in identifying cases of perseveration, we must distinguish between a developmentally immature system and, possibly, a system that is completely absent. In the case of adult cotton-top tamarins, the most likely explanation for their perseverative errors is the lack of an appropriate inhibitory mechanism. Third, although the locus of inhibitory control in this task appears to be the dorsolateral prefrontal cortex for humans and rhesus monkeys, research on common marmosets suggests that for this species—and most likely the closely related cotton-top tamarin as well—the orbitofrontal cortex is more important. This difference in functional anatomy is important because it suggests that over evolutionary time there may have been a shift in the locus of control for inhibitory mechanisms. Fourth, although the tamarins tested in group one perseverated with a straight-ahead response bias, the tamarins in group two did not. The latter group's performance suggests that when animals are given an alternative motor routine, or more colloquially "an alternative way of thinking about the problem," they can override the motivational drive. The fact that tamarins may not be capable of independently deriving an alternative response in this task suggests that one source of difficulty is conceptual as opposed to emotional.

Two Types of Perseveration: Cartesian and Kuhnian

For ease of discussion, I have proposed elsewhere (Hauser 1999, 2000) that we make a distinction between at least two types of perseverative error. On the one hand are cases where a response bias appears mediated by a prepotent emotion or motivational drive. I refer to these as cases of Cartesian perseveration, to reflect Descartes' intuition that animals, lacking rational thought, are driven by their passions. On the other hand are cases where a response bias appears mediated by an inappropriate conceptual framework. Consequently, the perseverative error arises because an individual fails to make a conceptual change. I refer to these as cases of Kuhnian perseveration to reflect Thomas Kuhn's (1970) insights into the nature of scientific change. Susan Carey (1985) has developed an analogous intuition with respect to the child's changing conceptual framework (see also Gopnik and Meltzoff 1997).[1] To illustrate the significance of this distinction, I first describe a case that I believe constitutes a pure example of Kuhnian perseveration and then discuss two additional cases that have elements of both types of perseveration.

Pure Kuhnian Perseveration

In several domains of knowledge, from naive physics to psychology, it appears that humans are endowed with conceptual biases. For example, as Spelke (1994) has articulated, human infants appear to be born with knowledge about physical objects, expecting them to behave in specific ways. Thus, in studies using the preferential looking-time method (Baillargeon 1995; Spelke 1985; Spelke et al. 1995), infants looked longer when objects failed to follow a continuous spatiotemporal path and when one stationary object moved off before being contacted by another (i.e., no action at a distance). These biases are important because they place the infant on a path of knowledge acquisition that is limited by the statistical regularities of the world. A critical problem then, is to document how conceptual change occurs within the child.

In a recent study by Hood (1995), 2.5–3.5-year-old children were presented with an invisible displacement task. The apparatus consisted of a rectangular frame with three chimneys (A, B, C) on top and three compartments (1,2,3) lined up directly below. While the child watched, an opaque tube was attached from one chimney (e.g., C) to one compartment (e.g., 1), thus creating an S-shaped configuration. A ball was then dropped down the chimney with the tube attached and the child was allowed to search for the ball in the compartments below. Consistently, children under the age of 3 looked in the compartment directly below the release point, even though they never found the ball in this compartment and always found it in the compartment with a tube attached. Moreover, when the tube was transparent, thereby allowing the child to watch the ball pass through and into the compartment, they succeeded in retrieving the ball, but as soon as the opaque tube was placed in the same configuration, they failed, going back to the compartment beneath the release point. Thus young children failed to generalize to the opaque tube, even though it was placed in exactly the same configuration as the transparent tube. Older children succeeded in finding the invisibly displaced ball even when the position of the tube changed from trial to trial, and even when a second distracter tube was inserted, thereby blocking the child's capacity to solve the problem by simple association (i.e., the ball is always in the compartment associated with a tube).

Based on these results, Hood argued that young children have a folk theoretic gravity bias, an expectation that causes them to search for invisibly displaced objects directly below the release point. Evidence for this claim comes from the robust observation that young children repeatedly search for the ball in the so-to-speak gravity compartment even in the face of counterevidence; that is, the ball never appears in the compartment below the release point. After age three, children apparently experience a conceptual shift, one that enables them to understand that the terminal location of a moving object depends in part on the objects in its path.

Further evidence for Hood's argument comes from a follow-up study. The tube apparatus was displayed on a computer monitor while an experimenter dropped a ball down the tube and then asked the child to point to the object's terminal location (Hood 1998). Replicating the original finding (Hood 1995), young children pointed to the compartment beneath the release point. However, when the apparatus was turned upside down, with compartments on top and chimneys down below, and the ball was sucked up the tubes, young children pointed to the correct compartment, that is, the one connected to the tube. This shows that the perseverative bias that arises when an object is invisibly displaced down along a vertical plane is not due to the child's inability to understand the physics of tubes. Rather, they seem to expect objects to fall straight down beneath the release point. As Hood puts it, young children seem to have a folk theory of gravity. Whether this bias is innately specified or acquired as a function of repeated exposure to falling objects is currently not clear. Nonetheless, young children must go through a conceptual revolution, overturning their beliefs about falling objects in order to include the possibility that some invisibly displaced objects are moved from their course by other objects.

To test whether this perseverative bias is due to an expectation or folk theory that is uniquely human, I teamed up with Hood and my two students, Linda Anderson and Laurie Santos, to run an exact replication of the original tube task with cotton-top tamarins (Hood et al. 1999). In the first trial, seven out of nine subjects searched for an invisibly displaced piece of food in the gravity compartment, and continued to search in this compartment over the course of several sessions (figure 11.3). In fact,

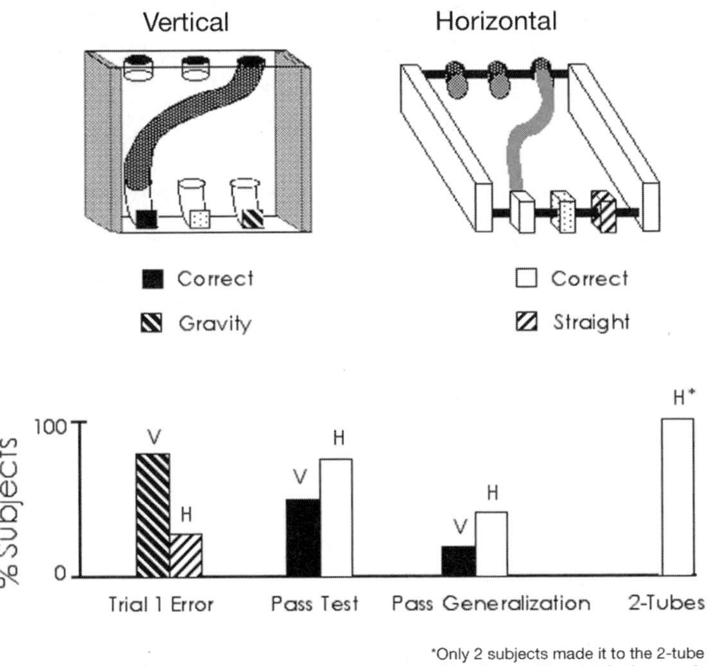

Figure 11.3
(Top) Apparatus for an invisible displacement problem for cotton-top tamarins. (Bottom) The results show that the tamarins performed significantly worse in the vertical than in the horizontal condition.

the perseverative bias was so strong in some of the tamarins that on a given trial they would repeatedly open the door to the compartment below the release point (i.e., no food) without opening any of the other compartments. The tamarins never solved this task with any consistency, even though they were given different conditions designed to help their performance. For example, like human children, they readily solved the problem involving a transparent tube, but then failed to generalize to the opaque tube on subsequent sessions. What is surprising about this failure, but also of considerable interest, is that the simplest solution to the vertical tube task is apparently unavailable to the tamarins; by straightforward association, the correct choice on each trial is always the compartment with a tube attached. Thus, even if a tamarin failed to watch the food drop into the tube, it could solve the problem by searching for the compartment associated with a tube. The tamarins failed to use this solution.

Because tamarins might have difficulty extracting the relevant information from a video display, we did not adopt Hood's video monitor procedure. Instead, we (Hauser et al. 2001) tested the tamarins on an equivalent task designed to assess whether their perseverative bias was due to difficulties with tubes or to invisible displacements more generally. We presented the tamarins with the tube apparatus positioned horizontally, thereby removing the forces of gravity (figure 11.3). Thus, instead of dropping a piece of food down the tube, we rolled it. In contrast to the task with vertical tubes, we observed no perseverative biases in the horizontal task; several subjects solved the single-tube task, generalized from transparent to opaque, and even solved the task when a distracter tube was put in place. Like Hood's inverted version of the vertical tubes, therefore, our horizontal task shows that the tamarins' poor performance in the original experiment (Hood et al. 1999) was not due to a difficulty with tubes. Rather, it appears that the perseverative bias in the original experiment was mediated by a powerful expectation, one that caused the tamarins to search for invisibly displaced objects along the vertical plane directly beneath the release point. That said, the subjects' performance on the horizontal task was not perfect, suggesting that tamarins may have more general difficulties with invisible displacements. To test this possibility and further explore the nature of the original perseverative error, we ran a third experiment (Hauser et al. 2001). This experiment was specifically set up to show whether the tamarins' experience on the tube tasks might help in subsequent problems involving invisible displacement, and whether their performance would improve if something other than a tube was responsible for deflecting an object from its path.

The apparatus was identical to the vertical tubes apparatus, except that we replaced the tube with an occluded ramp. Thus, the tamarins watched as we set down a rectangular frame with three notches on top (A, B, C) and three compartments (1, 2, 3) lined up below. We then attached a solid ramp from one notch (e.g., A) to one compartment (e.g., 3), and occluded the ramp with an opaque screen. When we dropped the food, eight out of twelve subjects searched in the compartment beneath the release point (e.g., compartment 1) on the first trial, and perseverated with this error over several trials. Surprisingly then, although most of the tamarins tested had experience with both the vertical and horizontal tubes, this experience apparently had no effect on their performance on the vertical ramps task. Moreover, even though this task involved ramps as opposed to tubes, the tamarins perseverated with the same bias, searching for the invisibly displaced food in the position directly beneath the release point. These data provide converging evidence that like 2.5–3.0-year-old human children, adult tamarins expect falling objects to fall straight down. Unlike human children, however, tamarins do not appear to be capable of performing the kind of conceptual change that is necessary to solve this problem, at least not in terms of the kind of experience that we have given them in these experiments and in their current living environment.

I view these experiments on children and tamarins as evidence of Kuhnian perseveration. That is, both young children and adult tamarins perseverate with a response bias because of a powerful conceptual framework, one that causes them to expect falling objects to land beneath the release point. Like other domains in which conceptual change has been investigated, conceptual frameworks are not overturned simply on the basis of accumulated facts (Carey 1985; Carey and Spelke 1994; Gopnik and Meltzoff 1997; Keil 1994). Although young children are given dozens and dozens of trials in which they fail to retrieve the ball in the compartment beneath the release point, they continue to search there because they firmly believe that this is the correct location. At some point, this theoretical expectation fails to provide a satisfactory explanation of their experiences and a conceptual change ensues—a minirevolution in the domain of physical knowledge.

Dissecting Perseveration: Kuhnian versus Cartesian Errors

Critical to any theory of morality is an understanding of why an individual perseverates with a particular response. If someone continues to steal, it is important to figure out whether this is because they are incapable of inhibiting their desires or whether they lack an understanding of the legal issues associated with stealing, and more deeply, the notion of fairness. In an attempt to explore this kind of problem from a cognitive neuroscience perspective, Damasio and his colleagues (Anderson et al. 1999; Bechara et al. 1997; Damasio 1994) have looked at patients with damage to the orbitofrontal cortex. In cases where the damage occurred in adulthood, the subjects were able to understand moral dilemmas and moral explanations for behavior, but had difficulty making decisions because they lacked the appropriate emotional responses. Consequently, they might perseverate with an inappropriate response because they lack the requisite emotional input that enables normals to properly evaluate a decision. In contrast, when damage to the orbitofrontal cortex occurs in infancy, a more dramatic conceptual deficit arises (Anderson et al. 1999). Specifically, based on tests of two patients, the results suggest that an understanding of moral dilemmas depends on the normal development of the orbitofrontal cortex. Consequently, when these two patients, who had been repeatedly convicted of petty crimes, were tested on Kohlberg's battery of moral problems (Colby and Kohlberg 1987), they scored in the range of a very young child. In brief, they simply failed to understand why something was wrong and why they should feel bad or good about their actions.

Studies of animals have not yet devised a battery of moral dilemmas (however, see the next section), but there are two experiments that reveal how Cartesian and Kuhnian perseverative errors might be dissected. In an ingenious yet simple experiment, Boysen (1996) presented two chimpanzees—a selector and a receiver—with a food choice task (figure 11.4). In each trial, one of two food wells always contained more food. The selector's task was to point to one food well. The receiver chimp obtained the food in the well pointed to, while the selector received the food in the other well. Thus, if chimpanzees are greedy—and we assume they are—then they should point to the well with less food. Over the course of dozens of trials in which selector and receiver changed roles and partners, no selector ever pointed to the smaller well with any consistency. That is, selectors most often pointed to the well with more food, and thus obtained less food. One interpretation of this finding is that chimpanzees are the ultimate altruists. An alternative interpretation is that chimpanzees lack the capacity to inhibit their desire to immediately obtain the larger quantity of food.

To test these alternative explanations, Boysen covered the food wells with cards associated with an Arabic numeral. Since the chimpanzees playing this game knew the meaning of the Arabic numerals, they understood that a card with the number "4" on it represented four pieces of food, while a card with "1" on it represented one piece of food. On the first trial of this transfer test, the chimpanzees pointed to the card with the lower number and thereby obtained the larger amount of food. This shows that chimpanzees are greedy, and that they can comprehend the rules of this game: Pick the one you don't want. It also shows that when the food is directly in view, they are incapable of inhibiting their greed. Moreover, and somewhat surprisingly, Boysen has repeatedly tested the chimps on the original, no-card task, and the results are the same; they continue to point to the larger quantity of food. Thus, even though they clearly understand the rules of this game, they are incapable of transferring their understanding from one context to another. The chimpanzees' error on the original condition is a clear case of Cartesian perseveration.

To evaluate the claim that chimpanzees lack inhibitory control in the original "pick the one you don't want" task, Silberberg and Fujita (1996) conducted a modified version of this task with Japanese macaques (figure 11.4). The experiment was motivated by the fact that whichever well the chimpanzees select, they are always rewarded, sometimes with the larger quantity and sometimes with the smaller. Given

Game: Pick the one you don't want

CHIMPANZEE

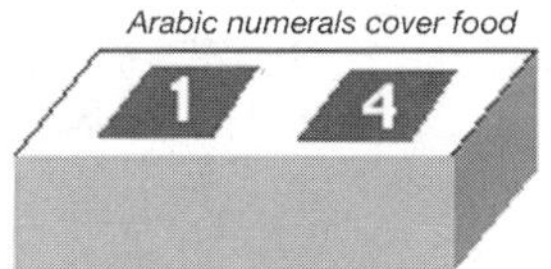

JAPANESE MACAQUE

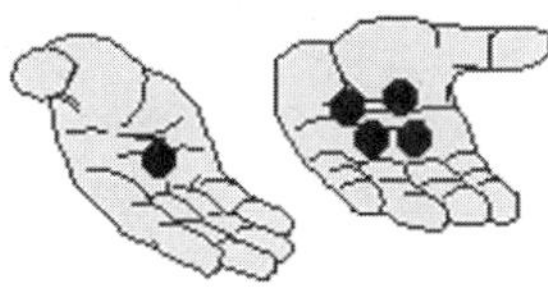

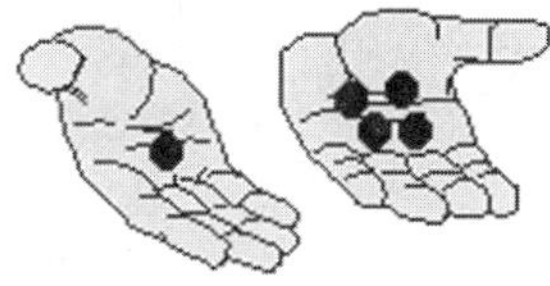

Figure 11.4

A test of reversed contingency learning involving a game with the rule: Pick the quantity you don't want, to get the quantity you want. Chimpanzees and Japanese macaques fail this task with food. When Arabic number cards replace the food, chimpanzees succeed on this task. When Japanese macaques incur a cost for picking the larger quantity (i.e., no food), they succeed as well.

that Boysen's chimpanzees are not food deprived, it is possible that the costs of picking the larger quantity are relatively trivial. This argument cannot account for the chimps' success with the Arabic number cards, but could account for their failure without these cards. In Silberberg and Fujita's experiment, they first tested the macaques on Boysen's original design and replicated the results exactly; the macaques consistently picked the well with more food. In a second condition, they changed the contingencies. Now, when the subjects selected the larger food quantity, they received no food at all. Under these conditions, the macaques switched strategies and consistently picked the well with less food. What this shows is that Japanese macaques, and presumably chimpanzees as well, can inhibit their desire to point to the well with more food if the costs of doing so are high. This does not negate the importance of Boysen's original finding, but does show how the problem of inhibition is a subtle one, and that it requires a careful investigation of an animal's motivational state.

In a second experiment, also designed to disentangle Cartesian and Kuhnian perseverative errors, cotton-top tamarins were presented with a means–end task involving the use of a tool to gain access to a piece of food (Hauser et al. 1999). In the original training condition, the subjects were presented with two potential tools. On one side was a piece of food on top of a piece of cloth. On the other side was an identical piece of food positioned next to the cloth, but not on top. To gain access to the food, a subject was required to pull the piece of cloth with the food on top; pulling the other cloth resulted in the cloth advancing, but without the food. The tamarins readily learned the solution to this problem and then generalized (on the first trial of each new condition) to a wide variety of novel conditions involving cloths of different color, texture, shape, and size, as well as connected and disconnected pieces (figure 11.5). These transfer trials show that the tamarins understand the means–end task. However, two additional conditions highlight the importance of challenging the conceptual abilities of animals with tasks that manipulate their motivational state. In one condition, the subjects were presented with two identical pieces of cloth. On one side a small piece of food was placed on top of the cloth while on the other side a large piece of food was placed off to the side of the cloth. Although the tamarins clearly understood the means–end task as demonstrated on hundreds of trials, in this condition they consistently reached for the cloth associated with the large piece of food; as a result, they perseverated with a response bias that failed to yield food. Apparently the desire to obtain a large piece of food overwhelmed their problem-solving abilities. A second condition reveals, however, the subtlety of the inhibitory problem. Here subjects were presented with two pieces of cloth with food located off the cloth. Thus, neither side of the tray provided the tamarins with an opportunity to gain food. In these trials, several tamarins looked at the tray and refrained from pulling either cloth. This shows that tamarins can inhibit cloth pulling even when there is food in view.

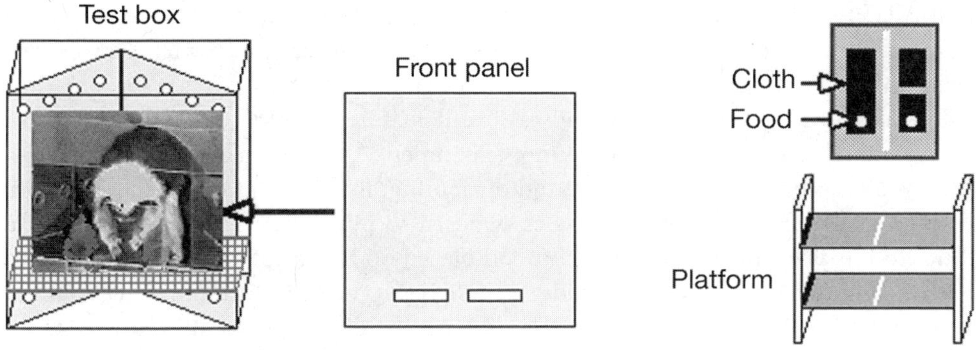

A sample of trials for tests of "On" and "Connected"

Affective challenges

Figure 11.5
A test of means-end problem-solving in cotton-top tamarins. The task involved pulling a piece of cloth with food located on top while avoiding cloths with a gap and cloths with food located off to the side. Tamarins solved this task, but when faced with some affective challenges (e.g., a large piece of food off, and small piece on), they failed.

To take the problem of Cartesian perseveration further, we have tested a naive group of tamarins on a different version of the cloth task (Hauser et al. 2002). This experiment is based on the results from other studies in our lab showing that for tamarins, shape and size represent functionally relevant features of tools, whereas color and texture do not (Hauser 1997). In the new experiment, the same means–end cloth conditions are set up, but in the first round of training, it just so happens that all of the correct responses are for blue cloths as opposed to pink ones. Thus, for example, a blue cloth is presented with a piece of food placed on top, as opposed to a pink cloth with a piece of food placed on the side. Although the correct solution involves pulling the cloth with the food located on top, pulling the blue cloth is also correct. Once the subjects reached criterion in this condition, they were transferred to a new condition where the correct color was now pink, but the means–end task was the same.

If tamarins approach this problem while thinking that the relevant problem has to do with tools, then based on our earlier work, they should ignore color and attend to shape (Hauser 1997; Hauser et al. 1999). However, if they treat this as a straightforward means–end task, where color provides the relevant clue, then in the transfer condition they should have great difficulty. The reason they will have problems is because by reversing the colors, we are effectively presenting the tamarins with a problem of reversal learning, a task that we know causes great difficulty for tamarins and many other primates. Results indicate that tool-inexperienced subjects have much greater difficulty solving this problem because during the first training condition they attended exclusively to color; the tool-experienced subjects solved the problem much faster, apparently because they either ignored color in the original training condition or attended to both color and functionality (shape).

In summary, I hope it is clear that in thinking about the problem of perseveration and the role of inhibitory mechanisms we must be careful to distinguish among the possible sources of such problems. Specifically, I have argued that there are at least two different kinds of perseverative error that may well rely on different neural circuitry: Cartesian and Kuhnian perseveration. Other researchers working in this general area have alluded to other classes of perseverative error, such as those that arise as a result of overlearning and attention (Dias et al. 1996; Hauser 1999). Future research on human and nonhuman animals, young and old, must be based on a battery of tests using both behavioral and neurobiological assays.

Do the Right Thing

David Copperfield was frequently forced to inhibit his most passionate desires. He of course would have been delighted to tell his stepfather Murdstone off, to drop his classes at the boarding school with the tyrannical schoolmaster Mr. Creakle, and to warn Em'ly that she was in danger of being snatched by the charms of unwanted lover

James Steerforth. But David squelched such desires, at least in terms of explicit action. Why did he squelch these desires? Was his ability to put on the brakes simply a reflection of an unconscious inhibitory mechanism? Or was there an intervening variable, something that told him that he was doing the right thing, acting in accordance with a set of moral guidelines? It is my contention that inhibitory mechanisms have been selected for because they guide individuals to do the right thing in the right context by preventing them from doing the wrong thing. The challenge we face in evaluating this hypothesis is to determine whether individuals who inhibit one action and favor another are doing so automatically or by recourse to a set of values that establish what is right and what is wrong. In this final section, I examine a set of experiments that were designed to challenge animals to do the right thing. In one situation, an animal is required to incur a personal cost so that another may avoid pain, a significant cost. In a second situation, an animal is required to incur a personal cost so that another may gain a benefit. Although these studies are a far cry from what many would consider to be relevant to our understanding of morality and its evolutionary origins, I find them compelling in terms of thinking about the requisite moral ingredients.

Helping Another in Pain

When one individual performs an act for the well-being or good of another, we call this act altruistic. Since the late 1960s and early 1970s, sociobiologists have argued that most altruistic acts are actually selfish (Dawkins 1976; Hamilton 1964; Trivers 1972; Williams 1966; E. O. Wilson 1975). This is because most altruism is directed toward close kin or unrelated individuals who are likely to pay back the favor. Consequently, if I help someone with whom I share genes in common, I am helping myself because I am functionally contributing to the replication of my genes. Similarly, if I help someone who will help me in the future, then I am acting in my own self-interest. Much less often are there clear-cut cases, at least among nonhuman animals, where altruism occurs among nonrelatives with no expectation of reciprocation (de Waal 1996). There are, however, a few candidate cases, and these come from experiments conducted in captivity. Although each of these experiments can readily be interpreted in more than one way, I find them interesting because they at least allow us to think about the problem of value-laden decisions in a comparative, evolutionary framework.

The animal psychologist Russ Church (1959) designed an experiment to determine whether genetically unrelated rats would act altruistically. He first trained a rat to press a lever for food, and then he changed the consequences of pressing the lever. On some trials, pressing the lever resulted in food; on other trials it delivered a shock to a second rat visible in an adjacent chamber. In the latter case, the rats in control of the lever immediately stopped pressing, thereby forfeiting the opportunity to eat, but saving another rat from shock. After several days of abstinence, some rats started pressing the lever, but did so at a lower rate than when no shock was delivered to the rat next door.

Do rats stop pressing because they can see that they are responsible for inflicting pain on another? Do they appreciate the causal connection between their lever pressing and the behavioral indicators of pain in another rat? To address this question, a second experiment was conducted. An experimenter trained a control group of rats to press for food. One test group pressed for food but occasionally received a shock. A second test group pressed for food but sometimes received a brief shock, while another visible rat received an extended shock. Relative to the control group, the first test group's response rate dropped by about 20 percent, while the second test group stopped responding altogether. Ten days later, at the end of the experiment, the second test group responded less than either of the other groups.

These experiments suggest that rats will eat less if, by inhibiting particular responses, another rat benefits. This looks like true altruism, based on sympathy or empathy toward another in pain. But is it? Perhaps rats stop pressing the lever because they find it particularly annoying and aversive to hear another rat's screaming and wriggling. All animals will seek ways to turn an unpleasant situation into something pleasant or at least less unpleasant. Moreover, the rat's ability to inhibit pressing generally fades over time, even though by pressing, it continues to deliver the shock. In brief, the rat's selfishness seems to take over. Is this not a reasonable response? Wouldn't we start pressing as well if we became hungry?

To determine whether rats would reduce another's potential suffering even when they gained no direct benefits, Rice and Gainer (1962) first trained one group to press a bar upon seeing a light; a shock was delivered if the rat failed to press the bar. In phase 2, if rats pressed the bar while the light was on, a Styrofoam block was lowered to the ground, left there for 15 seconds, and then hoisted back up; again, failure to press the bar resulted in shock. In phase 3, Rice and Gainer eliminated the shock, thereby eliminating the rats' motive to press the bar. These rats were then divided into two groups. The experimenter presented group A with the light and Styrofoam block hoisted above, and group B with the light and a rat suspended in the same position as the Styrofoam block; the rat squealed and wriggled while it was suspended. The results showed that the rats in group A almost never pressed, while the rats in group B pressed a lot; by pressing, they lowered the suspended rat to the ground and thereby functionally stopped his wriggling and screaming. This pattern of response suggests that the rats in group B were acting altruistically, providing a benefit to a genetically unrelated rat while incurring some cost and no direct personal benefit. Nonetheless, this experiment leaves us with the same kinds of questions we raised earlier in discussing Church's experiments. Did the rats in group B increase their rate of pressing because they wanted to relieve the suspended rat from distress? Have rats evolved a mechanism that selects for doing the right thing, helping another in need even though there is no direct benefit? Or did they press because the sound and sight of another rat screaming and wriggling is aversive?

To determine whether rats bar press to reduce another's suffering—an altruistic act—or to reduce the unpleasantness of another's noises—a selfish act—an experimenter presented one group of rats with white noise and a second group with recordings of rat squeals (Lavery and Foley 1963). If the subject pressed a bar, this terminated the sound playback. After several trials, the groups switched conditions, so that the second group heard white noise and the first group heard rat squeals. The results showed that both groups pressed the bar more in response to white noise. This suggests that white noise is more aversive than rat squeals. This fact, however, does not allow us to reject the idea that pressing the bar to rat squeals—or to the sight of a squealing rat—is altruistic. Rather, this result shows that rats will act to eliminate a variety of aversive stimuli, including rat squeals. They do the right thing, but we don't know why.

Rats might just be the wrong species for exploring whether nonhuman animals act in order to do the right thing. After all, rats lack the kind of social complexity that one might think is necessary for such morally guided actions. Acting with the intent to do the right thing is presumably selected for when there are societal pressures that would favor following the rules. If rats lack the requisite social complexity, then surely primates have it, especially the highly social Old World monkeys and apes.

In the 1960s, Robert Miller and his colleagues (Miller 1967; Miller et al. 1966, 1962, 1963) initiated a series of experiments to explore the nature of cooperative interaction in captive rhesus monkeys. Each experiment was designed to assess the conditions under which one animal might help another. In one of the first experiments, an experimenter trained two rhesus monkeys to avoid shock by pulling a lever in response to hearing a sound. On the test day, one of these individuals played the role of actor and was placed in a room with a lever and video image of the second animal, the receiver, located out of sight and hearing range. The receiver heard the sound played back during training—the one associated with shock—but did not have access to a response lever. What Miller and colleagues assumed is that when the receiver hears the sound and thus anticipates a shock, he will respond with a facial expression, one denoting fear. If the actor is able to read the receiver's facial expression, then he should use this information to time his response. He should press the lever in response to the receiver's facial expression, thereby avoiding a shock; the actor should be motivated to do this on selfish grounds alone, because failure to press resulted in shock to both the actor and the receiver. Since shock trials were presented randomly, and since neither animal could hear the another, there was no way to predict the timing of a response except by using the receiver's image in the monitor.

The actor pulled the lever significantly more when the receiver heard the sound than during silent periods. This shows that the actor used the receiver's facial expression to guide the timing of his response. Miller and his colleagues suggest that both animals had acquired a cooperative response; to avoid shock, the receiver must signal some information and the actor must read the signals.

To conclude, as Miller and colleagues do, that the receiver and actor cooperated, one has to assume that the receiver intended to provide information to the actor, as opposed to incidentally providing information. I don't believe this assumption is warranted. The receiver must have felt helpless, distressed, and afraid. To say that the receiver was signaling, however, one would have to show that he knew about the actor's presence. Given the experimental setup, he most certainly did not. Rather, the receiver's response was elicited by the sound, perhaps as reflexively as we kick out our knee in response to the doctor's tiny mallet. Furthermore, it seems more likely that the actor picked up on a change in the receiver's activity, one that was sufficiently consistent to be an accurate predictor of things to come. Using an expression to predict a response is not the same as viewing the expression as an indication of another's emotions.

This experiment is on the right track, given our concerns with morally guided actions, but it also leaves many loose ends. Although it is clear that rhesus monkeys can avoid shock by watching another-monkey's facial expressions, we do not know whether this response is mediated by empathy, an emotional capacity that is necessary for true altruism. One has to feel what it would be like to be someone else, to feel fear, pain, or joy. We don't know whether the actor was even aware of the receiver's feeling. And to be perfectly blunt, there is no reason for the actor to care. All that matters to the actor is that the image displayed on the monitor reliably predicts shock. A better experiment would allow the actor to see what was happening to the receiver, but would restrict the shock to the receiver.

To push the issue further, a different experiment was run, playing off the findings from the previous work (Wechkin et al. 1964). Several rhesus monkey actors were trained to pull one of two chains to receive food. This was the only source of food for them on a given day and thus they were highly motivated to pull. Next, an actor was paired with a second rhesus monkey—the receiver—who was introduced into an adjacent cage, allowing both animals to see each other. In the second phase, the contingencies associated with pulling a chain changed. One chain continued to deliver food, whereas the second delivered a severe shock to the receiver. The results showed that most actors reduced the number of responses to the chain delivering the shock, especially when contrasted with the chain delivering food. One subject stopped pulling both chains for 5 days, and another for 12 days. This self-starvation (i.e., no pulling, no food for the day) happened more in animals who had experienced shocks themselves. When the actors were paired up with new receivers, most continued to refrain from pulling the chain delivering the shock, and actors withheld from pulling for longer when the receiver was a familiar cage mate. It's as if altruism was affected by friendship, shifting the economics in favor of helping another in need if the other is well known.

Although one must clearly acknowledge the abhorrent nature of these experiments, in that delivery of shock is inhumane, as is the psychological consequences of putting animals in a situation to shock another or starve, we can learn from such experiments. Specifically, what I find most remarkable about the results from these experiments is the observation that some rhesus monkeys refrained from eating in order to avoid injuring another individual. Perhaps the actors empathized, feeling what it would be like to be shocked, what it would be like to be the other monkey in pain. Alternatively, perhaps seeing someone shocked is unpleasant, and rhesus monkeys will do whatever they can to avoid unpleasant conditions. Or, perhaps actors stopped pulling because they realized that one day it might be their turn to play the role of receiver. At this point, it is simply not possible to distinguish between an empathetically motivated altruistic act and a selfishly motivated one.

When to Cooperate

After a long and arduous childhood, the 10-year-old David Copperfield leaves the filth of a London warehouse and treks to visit his great-aunt, Betsey Trotwood. He hasn't seen her in ages, and all that he can recall about her is that she thinks David was born the wrong sex! Although she is initially appalled by his appearance, Betsey warms to David, and eventually they form an extraordinarily close bond. She realizes that he needs help, and even steps in to defend him against the horrid Mr. Murdstone who, during a brief visit, attempts to take David back. A cynical sociobiological perspective would say that Betsey has cooperated with David because they are genetically related. But as the reader will know, the genetic overlap between David and Betsey is actually quite small, and perhaps unlikely to account for her altruistic act. Nonetheless, as we learn later in the book, David returns the favor by helping Betsey with her troubled finances. In each case, both Betsey and David clearly do the right thing.

The previous section provided some evidence that animals may help to alleviate another's pain. Do they help by providing resources to others or by sharing the resources that they have gathered? If such altruistic actions arise, are they reciprocated? Although there is little experimental work on reciprocal altruism, especially with respect to identifying some of the underlying psychological mechanisms (see review in Hauser 2000), the few studies conducted to date are telling. Here I review two such studies.

What would happen in an animal society if one member acquired a skill that no other animal had? And what if this skill resulted in considerable benefits with respect to resource acquisition? Would others develop new respect for this skilled individual? Would they start cooperating in order to gain favors in the future? To explore this hypothetical situation, the ethologist Eduard Stammbach (1988) set up an experiment with long-tailed macaques (*Macaca fascicularis*). All members of several social groups were trained to press a lever for popcorn whenever a light was illuminated. Then the lowest-ranking member in each group was trained to press a set of levers in a specific sequence to deliver enough popcorn for three individuals.

The results were as follows: Initially, the higher-ranking individuals chased the skilled low ranking individuals away from the dispenser, thereby intimidating them. The higher-ranking individuals learned, however, that the low-ranking individual had a unique skill, and followed it to the machine, waiting to grab all of the popcorn. As a result, low-ranking specialists stopped working the machine. Soon thereafter, some higher-ranking individuals stopped chasing specialists away from the dispenser. Instead, when the skilled individual approached the dispenser and worked the levers, high-ranking members of the group inhibited all aggression, approached peacefully, and shared in the popcorn bounty. Furthermore, high-ranking individuals started grooming the specialists more often, even during periods when the machine was inoperative. Although this change by the dominants enabled low-ranking specialists to access food that would normally be unobtainable, it had no impact on their dominance rank within the group. The specialists kept their low rank but were allowed brief moments with the dominant elite when their skills were of use.

What is perhaps most striking about this experiment is the consistency with which nonspecialists responded to specialists. The specialist was tolerated at the popcorn dispenser independent of age, rank, or sex. Furthermore, nonspecialists appear to have recognized that access to popcorn from the dispenser required an active investment in the specialists, grooming them and suppressing a natural tendency to chase them away from a valued resource. As previously discussed, high-ranking group members had to inhibit their desire to attack low-ranking specialists over access to food; they also had to provide an incentive for the specialist to operate the dispenser. Clearly, high-ranking animals can inhibit their emotions in this context, as evidenced by the decrease in aggression toward specialists and their toleration of specialists at the popcorn dispenser.

Is it reasonable to conclude that high-ranking macaques learn to respect specialists, placing values on their unique skills, and then cooperating with them? Do nonspecialists tolerate specialists at the popcorn dispenser because they think it is fair? Or is toleration the result of self-interest, the realization that the only route to obtaining popcorn is to allow specialists to share in the food they have worked to obtain. At present, we cannot discriminate between these two alternative interpretations. To show that macaques, or any other organism, has a sense of fairness or respect, we would have to show that they have a sense of what is right and wrong, and that they act on the basis of such principles. At some level, then, we must be able to show that they adhere to particular rules of order, engaging in behaviors that maintain the peace and prevent chaos, war, and anarchy. They must not only act within a set of normative rules, but must accompany such actions with a set of normative feelings.

A recent experiment by Frans de Waal and Michelle Berger (2000) looks at the economics of cooperation in more detail by providing capuchin monkeys with an opportunity to share food. There were three conditions, each involving two subjects, and an apparatus that provided access to food. In one condition (solo), one monkey had access to a bar attached to a bowl with food; pulling the bar brought the bowl closer and allowed the actor access to food. In a second condition (cooperation), there were two bowls and two pull bars, but only one bowl was filled with food; in order to bring the bowls closer, both subjects had to pull the bars. The third condition (mutualism) was the same as the second, except both bowls were filled with food. Capuchins were more successful in pulling in the food bowls for the solo and cooperation conditions than for the mutualism condition. Since capuchins will often allow others to take food from them, a behavior called "facilitated taking" by de Waal and Berger, or "tolerated theft" by Blurton-Jones (1987), it was possible to look at the amount of food obtained by the passive subject in the solo condition and by the subject (i.e., the helper) with an empty food bowl in the cooperation condition. The results showed that capuchins were more likely to allow facilitated taking in the cooperation condition than in the solo condition. Moreover, helpers were more likely to pull on a cooperation trial if the previous cooperation trial was successful (i.e., yielding food for both).

De Waal and Berger's results are elegant, showing that capuchins, and presumably some other animals as well, will allow others to share in the food obtained if they have received help in securing it. Moreover, individuals appear to keep track of cooperative interactions, using the economics of the outcome to determine what happens in the future. As de Waal and Berger conclude, "The increase in sharing following cooperation may rest on psychological mechanisms as complex as mental score-keeping of services and 'gratitude,' or as simple as attitudinal reciprocity. According to the latter explanation, a joint effort, and the mutual coordination this entails, may induce a positive attitude towards the partner, reflected in attraction and social tolerance. If this facilitates the sharing of pay-offs, in turn providing an incentive for continued cooperation, we have two mechanisms that together function as payment for labour and labour for payment" (2000: 563). There is no doubt that these represent two plausible mechanisms. The challenge for future research is to develop new experiments that focus explicitly on the underlying psychological mechanisms and determine whether the capacity for cooperation exhibited by capuchins is present in other primate and nonprimate species.

How to Do the Right Thing

In this chapter I have argued that a first step in understanding the evolution of moral systems is to break morality down into its most substantive ingredients. In this sense I follow the writings of Kant and Darwin. To illustrate the promise of this approach, I have focused on one, albeit powerful ingredient: the capacity to inhibit one action and favor others. Inhibition is not, however, a simple character trait. Rather, there are different kinds of inhibition, and these must be identified in our attempts to discern how they evolved across species and how they develop within a species. In some cases, nonhuman animals show a remarkable lack of inhibitory control, which leads to perseverative errors in problems involving motivational and conceptual challenges. In other cases, animals show great restraint, forfeiting access to food to alleviate another's pain or to help another who needs food. Whether they act in these ways with the intent to do the right thing is another matter. Ultimately, therefore, it will be necessary to explore whether an animal that inhibits an action in order to benefit another does so while thinking about the other's emotional or mental states. Although nature may not be replete with the moral fibers of a David Copperfield, the ingredients from which Copperfield developed, and from which his species evolved, may be right in front of our eyes. We simply must look, observe, and carry out the *right* experiments.

Note

1. Apologies to Susan Carey for not naming this Carey-ian perseveration, but for historical reasons, Kuhn was there first!

References

Alexander, R. D. 1987. *The biology of morality systems*. New York: Aldine.

Anderson, S. W., Bechara, A., Damasio, H., Tranel, D., and Damasio, A. R. 1999. Impairment of social and moral behavior related to early damage in human prefrontal cortex. *Nature Neuroscience* 2: 1032–1037.

Baillargeon, R. 1995. Physical reasoning in infancy. In M. Gazzaniga (ed.), *The cognitive neurosciences* (pp. 181–204). Cambridge, Mass.: MIT Press.

Bechara, A., Damasio, H., Tranel, D., and Damasio, A. R. 1997. Deciding advantageously before knowing the advantageous strategy. *Science* 275: 1293–1294.

Blurton-Jones, N. 1987. Tolerated theft, suggestions about the ecology of sharing, hoarding, and scrounging. *Biology and Social Life* 26: 31–54.

Boysen, S. T. 1996. "More is less": The distribution of rule-governed resource distribution in chimpanzees. In A. E. Russon, K. A. Bard, and S. T. Parker (eds.), *Reaching into thought: The minds of the great apes* (pp. 177–189). Cambridge: Cambridge University Press.

Carey, S. 1985. *Conceptual change in childhood*. Cambridge, Mass.: MIT Press.

Carey, S., and Spelke, E. S. 1994. Domain-specific knowledge and conceptual change. In L. Herschfeld and S. Gelman (eds.), *Mapping the mind: Domain-specificity in cognition and culture* (pp. 169–201). Cambridge: Cambridge University Press.

Chomsky, N. 1957. *Syntactic structures.* The Hague: Mouton.

Chomsky, N. 1986. *Knowledge of language: Its nature, origin, and use.* New York: Praeger.

Church, R. M. 1959. Emotional reactions of rats to the pain of others. *Journal of Comparative and Physiological Psychology* 52: 132–134.

Colby, A., and Kohlberg, L. 1987. *The measurement of moral judgment.* New York: Cambridge University Press.

Damasio, A. 1994. *Descartes' error.* New York: Putnam.

Darwin, C. 1871. *The descent of man and selection in relation to sex.* London: John Murray.

Darwin, C. 1872. *The expression of the emotions in man and animals.* London: John Murray.

Dawkins, R. 1976. *The selfish gene.* Oxford: Oxford University Press.

Diamond, A., 1988. Differences between adult and infant cognition: Is the crucial variable presence or absence of language? In L. Weiskrantz (ed.), *Thought without language* (pp. 337–370). Oxford: Clarendon Press.

Diamond, A., 1990. Developmental time course in human infants and infant monkeys, and the neural bases of higher cognitive functions. *Annals of the New York Academy of Sciences* 608: 637–676.

Diamond, A., and Goldman-Rakic, P. S. 1989. Comparison of human infants and infant rhesus monkeys on Piaget's AB task: Evidence for dependence on dorsolateral prefrontal cortex. *Experimental Brain Research* 74: 24–40.

Diamond, A., Zola-Morgan, S., and Squire, L. R. 1989. Successful performance by monkeys with lesions of the hippocampal formation on AB and object retrieval, two tasks that mark developmental changes in human infants. *Behavioral Neuroscience* 103: 526–537.

Dias, R., Robbins, T. W., and Roberts, A. C. 1996. Dissociation in prefrontal cortex of affective and attentional shifts. *Nature* 380: 69–72.

Gopnik, A., and Meltzoff, A. 1997. *Words, thoughts, and theories.* Cambridge, Mass.: MIT Press.

Hamilton, W. D. 1964. The evolution of altruistic behavior. *American Naturalist* 97: 354–356.

Hauser, M. D. 1997. Artifactual kinds and functional design features: What a primate understands without language. *Cognition* 64: 285–308.

Hauser, M. D. 1999. Perseveration, inhibition, and the prefrontal cortex: A new look? *Current Opinions in Neurobiology* 9: 214–222.

Hauser, M. D. 2000. *Wild minds: What animals really think.* New York: Henry Holt.

Hauser, M. D., Kralik, J., and Botto-Mahan, C. 1999. Problem solving and functional design features: Experiments with cotton-top tamarins. *Animal Behaviour* 57: 565–582.

Hauser, M. D., Williams, T., Kralik, J. D., and Moskovitz, D. (2001). What guides a search for food that has disappeared? Experiments on cotton-top tamarins (*Saguinus oedipus*). *Journal of Comparative Psychology* 115(2): 140–151.

Hauser, M. D., Santos, L., Spaepen, G., and Pearson, H. E. (2002). Problem-solving, inhibition, and domain-specific experience. *Animal Behaviour* 64: 387–396.

Hood, B. M. 1995. Gravity rules for 2–4 year olds? *Cognitive Development* 10: 577–598.

Hood, B. M. 1998. Gravity does rule for falling objects. *Developmental Science* 1: 59–64.

Hood, B. M., Hauser, M. D., Anderson, L., and Santos, L. 1999. Gravity biases in a nonhuman primate? *Developmental Science* 2: 35–41.

Jackendoff, R. 1994. *Patterns in the mind.* New York: Basic Books.

Kant, I. 1781(1929). *Critique of pure reason.* London: Macmillan.

Kant, I. 1797(1965). *The metaphysical elements of justice.* Indianapolis, Ind.: Bobbs-Merrill.

Keil, F. C. 1994. The birth and nurturance of concepts by domains: The origins of concepts of living things. In L. A. Hirschfeld and S. A. Gelman (eds.), *Mapping the mind: Domain-specificity in cognition and culture* (pp. 234–254). Cambridge: Cambridge University Press.

Kuhn, T. S. 1970. *The structure of scientific revolutions.* Chicago: University of Chicago Press.

Lavery, J. J., and Foley, P. J. 1963. Altruism or arousal in the rat? *Science* 140: 172–173.

Miller, R. E. 1967. Experimental approaches to the physiological and behavioral concomitants of affective communication in rhesus monkeys. In S. A. Altmann (ed.), *Social communication among primates* (pp. 125–135) Chicago: University of Chicago Press.

Miller, R. E., Banks, J., and Ogawa, N. 1962. Communication of affect in "cooperative conditioning" of rhesus monkeys. *Journal of Abnormal Social Psychology* 64: 343–348.

Miller, R. E., Banks, J., and Ogawa, N. 1963. The role of facial expression in "cooperative-avoidance" conditioning in monkeys. *Journal of Abnormal Social Psychology* 67: 24–30.

Miller, R. E., Banks, J., and Kuwahara, H. 1966. The communication of affects in monkeys: Cooperative conditioning. *Journal of Genetic Psychology* 108: 121–134.

Piaget, J. 1952. *The origins of intelligence in children.* New York: International University Press.

Pinker, S. 1994. *The language instinct.* New York: William Morrow.

Preuss, T. M. 1995. The argument from animals to humans in cognitive neuroscience. In M. Gazzaniga (ed.), *The cognitive neurosciences* (pp. 1227–1243). Cambridge, Mass.: MIT Press.

Rice, G. E., and Gainer, P. 1962. "Altruism" in the albino rat. *Journal of Comparative and Physiological Psychology* 55: 123–125.

Ridley, M. 1996. *The origins of virtue.* New York: Viking Press/Penguin Books.

Roberts, A. C., and Wallis, J. D. 2000. Inhibitory control and affective processing in the prefrontal cortex: Neuropsychological studies in the common marmoset. *Cerebral Cortex* 10: 252–262.

Roberts, A. C., Robbins, T., and Weiskrantz, L. 1998. *The prefrontal cortex: Executive and cognitive functions.* Oxford: Oxford University Press.

Santos, L. R., Ericson, B., and Hauser, M. D. 1999. Constraints on problem solving and inhibition: Object retrieval in cotton-top tamarins. *Journal of Comparative Psychology* 113: 1–8.

Silberberg, A., and Fujita, K. 1996. Pointing at smaller food amounts in an analogue of Boysen and Bernston's (1995) procedure. *Quarterly Journal of Experimental Psychology* 66: 143–147.

Sober, E., and Wilson, D. S. 1998. *Unto others.* Cambridge, Mass.: Harvard University Press.

Spelke, E. S. 1985. Preferential looking methods as tools for the study of cognition in infancy. In G. Gottlieb and N. Krasnegor (eds.), *Measurement of audition and vision in the first year of postnatal life* (pp. 85–168). Norwood, N.J.: Ablex.

Spelke, E. S. 1994. Initial knowledge: Six suggestions. *Cognition* 50: 431–445.

Spelke, E. S., Vishton, P., and von Hofsten, C. 1995. Object perception, object-directed action, and physical knowledge in infancy. In M. Gazzaniga (ed.), *The cognitive neurosciences* (pp. 165–179). Cambridge, Mass.: MIT Press.

Stammbach, E. 1988. Group responses to specially skilled individuals in a *Macaca fascicularis* group. *Behavior* 107: 241–266.

Trivers, R. L. 1972. Parental investment and sexual selection. In B. Campbell (ed.), *Sexual selection and the descent of man* (pp. 136–179). Chicago: Aldine.

de Waal, F. B. M. 1996. *Good natured.* Cambridge, Mass.: Harvard University Press.

de Waal, F. B. M., and Berger, M. L. 2000. Payment for labour in monkeys. *Nature* 404: 563.

Wechkin, S., Masserman, J. H., and Terris, W., Jr. 1964. Shock to a conspecific as an aversive stimulus. *Psychonomic Science* 1: 47–48.

Williams, G. C. 1966. *Adaptation and natural selection.* Princeton, N.J.: Princeton University Press.

Wilson, E. O. 1975. *Sociobiology.* Cambridge, Mass.: Harvard University Press.

Wilson, J. Q. 1993. *The moral sense.* New York: Free Press.

Wright, R. 1994. *The moral animal.* New York: Pantheon.

12 The Cultural and Evolutionary History of the Real Numbers[1]

C. R. Gallistel, Rochel Gelman, and Sara Cordes

The cultural history of the real numbers began with the positive integers. Kronecker is often quoted as saying, "God made the integers; all else is the work of man," by which he meant that the system of real numbers had been erected by mathematicians on the intuitively obvious foundation provided by the integers. Taken as a statement about the cultural history of mathematics, this is beyond dispute. However, if this is taken as a claim about the psychological foundations of arithmetic reasoning, then we suggest that here, as in many other areas of psychology, introspection and intuition are poor guides to the inner workings of the mind.

We suggest that it is the system of real numbers that is the psychologically primitive system, both in the phylogenetic and the ontogenetic sense. We review evidence that a system for arithmetic reasoning with real numbers evolved before language evolved. When language evolved, it picked out from the real numbers only the integers, thereby making the integers the foundation of the cultural history of the number. Second, we suggest that this ancestral nonverbal real number system becomes operative in the prelinguistic child and makes possible the acquisition of language-mediated counting and language-mediated arithmetic reasoning. It is the foundation on which an individual's language-mediated understanding of what numbers are and what may be done with them rests.

The Formal Relation Between the Integers and the Reals

The number system that can be used to represent continuous (uncountable) quantities is the system of real numbers. It includes the irrational numbers, such as $\sqrt{2}$, and the transcendental numbers, such as π. It is used by modern humans to represent many distinct systems of continuous quantity—duration, length, area, volume, density, rate, intensity, and so on. Because the system of real numbers is isomorphic to a system of magnitudes, the terms *real number* and *magnitude* are used interchangeably. Thus, when we refer to mental magnitudes, we are referring to a real number system in the brain. Like the culturally specified real number system, the real number system in the brain is used to represent both continuous quantity and numerosity.

Magnitudes and real numbers have the property that there is no way to pick out a successor, the next number in the sequence. Given a line of some length, there is no procedure by which one can pick out the next longer line. Similarly, given a real number, such as, say, 2, there is no procedure that picks out the next real number, although there is of course a procedure that picks out the next integer. The real numbers are not discretely ordered, but the integers are, so 2 qua real number has no successor, whereas 2 qua integer does.

The discrete ordering of the natural numbers (the positive integers) makes them uniquely suited to represent numerosity, that is, countable quantity. The positive integers, however, taken by themselves, rather than as a component of the system of real numbers, have two serious failings—an algebraic failing and a geometric failing. The algebraic failing is that they are not closed under the inverse combinatorial operations of subtraction and division. Subtracting one positive integer from another often fails to yield a positive integer. If only the positive integers are regarded as legitimate numbers, then subtraction can be legitimately performed only when it is known to yield a positive integer. However, in the course of algebraic reasoning, it is often desirable to subtract one unknown number from another. If only positive integers are allowed as numbers, then this maneuver will be of doubtful legitimacy because one will not know whether the subtrahend is larger or smaller than the minuend. The division of one unknown number by another is likewise suspect because only rarely does dividing one positive integer by another yield a positive integer.

The lack of closure in the system of natural numbers provided much of the motivation that drove the cultural expansion of that system to include zero, the negative integers, and the rational numbers. The rational numbers include all the numbers that may be expressed as the proportion between two integers, that is, the fractions, including the improper fractions such as 71/53. The geometric failing of the integers, and their offspring the rational numbers, arises when we attempt to use proportions between integers to represent proportions between continuous quantities, as, for example, when we say that one person is half again as tall as another, or one farmer has only a tenth as much land as another. These locutions show the seemingly natural expansion of the integers to the rational numbers, numbers that represent proportions. This expansion seemed so natural and unproblematic to the Pythagoreans that they believed that the natural numbers and the proportions between them (the rational numbers) were the inner essence of reality, the carriers of all the deep truths about the world. They were, therefore, greatly unsettled when they discovered that there were geometric proportions that could not be represented by a rational number, for example, the proportion between the diagonal and the side of a square. The Greeks proved that no matter how fine you made your unit of length, it would never go an integer number of times into both the side and the diagonal. Put another way, they proved that the square root of 2 is an irrational number, an entity that cannot be constructed from the natural numbers by a finite procedure.

As the name they gave to these would-be numbers implies, the Greeks found the existence of irrational numbers contrary to reason. They sensed that number ought to be able to represent geometric proportions like this, and they were right. They were right because if we follow the impulse to create a closed algebraic system to its natural end, then we are led to the system of real numbers, indeed, eventually to the system of complex numbers. The system of real numbers has numbers to represent every geometric proportion. Thus, in the process of creating the numbers needed to guarantee algebraic closure, mathematicians created the numbers needed to represent every possible proportion. Our thesis is that this cultural creation of the real numbers was a Platonic rediscovery of the underlying nonverbal system of arithmetic reasoning. The cultural history of the number concept is the history of our learning to talk coherently about a system of reasoning with real numbers that predates our ability to talk, both phylogenetically and ontogenetically.

Evidence for the Ancestral Status of the Real Numbers

Other Vertebrates Measure and Remember Uncountable Magnitudes

The common laboratory animals, such as the pigeon, the rat, and the monkey, measure and remember continuous quantities, such as duration, as has been shown in a variety of experimental paradigms. One of these is the so-called peak procedure. In this procedure, a trial begins when a stimulus signaling the possible availability of food comes on. When pigeons are the subjects, the stimulus is the illumination of a key on the wall of the experimental chamber. When the subjects are rats, the stimulus is the extension into the cage of a lever. On 25–50 percent of the trials, the key or lever is armed at a fixed latency after the onset of the stimulus. Pecking or pressing before the key or lever is armed is pointless, but the first peck or press after the arming delivers food. On the remaining 75–50 percent of the trials, however, the key or lever is not armed. On these trials, the key remains illuminated or the lever remains extended for between four and six times longer than the arming latency. Pecking or pressing after the arming latency has elapsed is pointless, because if there has been no reward at that latency, then there will be none on that trial.

Peak-procedure data come from the unrewarded trials. On such trials, the subject abruptly begins to peck the key or press the lever some time before the arming latency (when it judges arming to be nigh) and continues to peck or press for some time afterward before abruptly stopping (when it judges that the arming latency has past). The interval during which the subject pecks or presses brackets its subjective estimate of the arming latency. Representative data for rats are shown in figure 12.1.

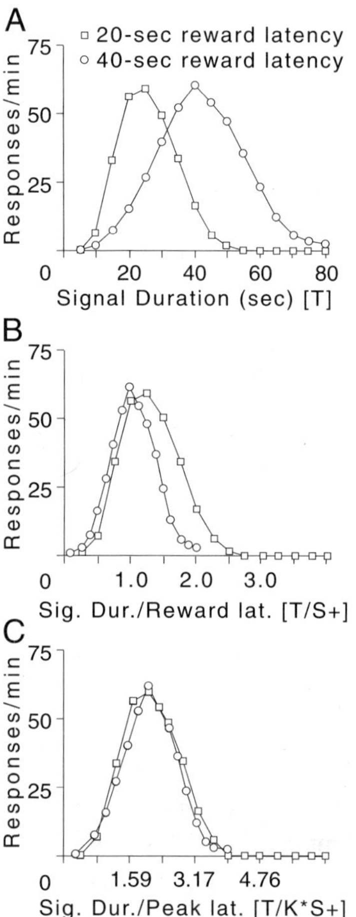

Figure 12.1

Representative peak-procedure data. The subjects were rats. In one block of many trials, the arming latency was 20 seconds; in another, it was 40 seconds. (A) The original data. (B) Data plotted as a proportion of the arming latency. (C) Data plotted as a proportion of the latency at the mode of the distributions in (A). Because the variability in the onsets and offsets of responding is proportional to the remembered arming latency, the distributions superimpose when they are plotted as a proportion of the modal latency.

Figure 12.1A, shows seemingly smooth increases and decreases in response rates on either side of the arming latency. The smoothness is an averaging artifact. On any one trial, the onset and offset of responding is abrupt. The temporal locus of these onsets and offsets varies from trial to trial. Averaging across trials gives these approximately normal distributions. The curves in figure 12.1A are best read as showing the probability that the subject will be responding as a function of the time elapsed since the warning signal came on. The mode of the distribution (the latency at which the distribution peaks) is the latency at which the subject is maximally likely to be responding. This latency does not necessarily coincide with the actual arming latency, because individual subjects often show small proportional errors in the mode; they misremember experienced durations by some multiplicative factor slightly greater or smaller than 1.

The distribution obtained with the 40-second arming latency is broader than the distribution obtained with the 20-second latency. As shown in figure 12.1C, the broadening of the distribution at longer arming latencies is proportional to the remembered arming durations (the mode of the distribution), not to the actual arming durations (figure 12.1B). When mean response rates are plotted against the elapsed proportion of the modal latency, the distributions obtained at different arming latencies superimpose (figure 12.1C). Thus, the trial-to-trial variability in the onsets and offsets of responding is proportional to the remembered latency. Put another way, the probabilities that the subject will have begun to respond or will have stopped responding are determined by the proportion of the remembered arming latency that has elapsed. This property of the memory for durations is called scalar variability.

Scalar variability is a ubiquitous property of remembered mental magnitudes. It seems to be best explained by the assumption that the neural signals that come from the reading of a memory show trial-to-trial (reading-to-reading) variability, just as do the neural signals that come from the action of a stimulus (Gallistel 1999; Gallistel and Gibbon 2000). In other words, the reading of a mental magnitude in memory is a noisy process, and the noise is proportional to the magnitude being read.

Other Vertebrates Count and Remember Numerosity

Rats, pigeons, and monkeys also count and remember numerosities (see Dehaene 1997; Gallistel 1990; Gallistel and Gelman 2000; for reviews see e.g., Roberts et al. 2000). One of the early protocols for assessing counting and numerical memory was developed by Mechner (1958) and later used by Platt and Johnson (1971). The subject must press a lever some number of times (the target number) in order to arm the infrared beam at the entrance to a feeding alcove. When the beam is armed, interrupting it releases food. Pressing too many times before trying the alcove incurs no penalty beyond that of having made supernumerary presses. Trying the alcove prematurely incurs a 10-second time-out, which the subject must endure before

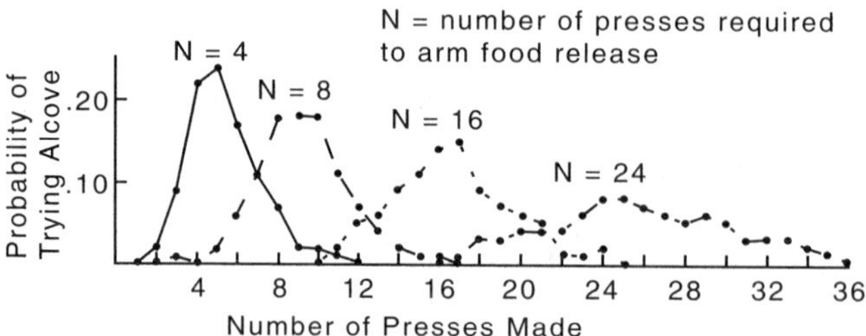

Figure 12.2
The probability of breaking off to try the feeding alcove as a function of the number of presses made on the arming lever and the number required to arm the food-release beam at the entrance to the feeding alcove. The subjects were rats. (Redrawn from Platt and Johnson, 1971 by permission of the authors and publishers.)

returning to the lever to complete the requisite number of presses. Data from such an experiment are shown in figure 12.2. They look strikingly like the temporal data. The number of presses at which subjects are maximally likely to break off pressing and try the alcove peaks at or slightly beyond the required number, for required numbers ranging from 4 to 24. As the remembered target number becomes larger, the variability in the break-off number also becomes proportionately greater. Thus, the memory for number also exhibits scalar variability.

The fact that the memory for numerosity exhibits scalar variability suggests that numerosity is represented in the brains of nonverbal vertebrates like rats, pigeons, and monkeys by mental magnitudes, that is, by real numbers, rather than by discrete symbols like words or bit patterns. When a device such as an analog computer represents numerosities by different voltage levels, noise in the voltages leads to confusion between nearby numbers. If, by contrast, a device represents countable quantity by countable (that is, discrete) symbols, as digital computers and written number systems do, then one does not expect to see the kind of variability seen in figure 12.2. For example, the bit-pattern symbol for 15 is 01111 whereas for 16 it is 10000. Although the numbers are adjacent, the discrete binary symbols for them differ in all five bits. Jitter in the bits (uncertainty about whether a given bit was 0 or 1) would make 14 (01110), 13 (01101), 11 (01011), and 7 (00111), all equally and maximally likely to be confused with 15 because the confusion arises in each case from the misreading of one bit. These dispersed numbers should be confused with 15 much more often than is the adjacent 16. Similarly, a scribe copying a handwritten English text is presumably more likely to confuse "seven" and "eleven" than to confuse "seven" and "eight." Thus, the nature of the variability in a remembered target number implies that what is being remembered is a magnitude—a real number.

Numerosity and duration are represented by comparable mental magnitudes. Meck and Church (1983) pointed out that the mental accumulator model that Gibbon

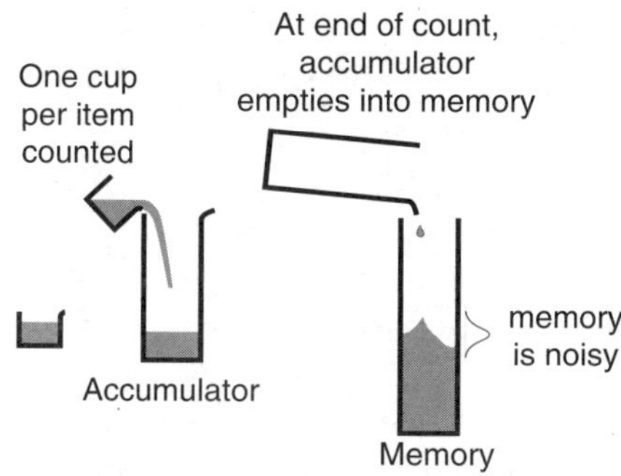

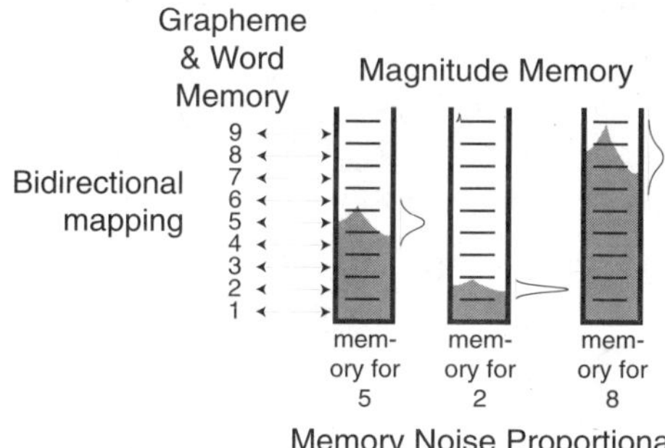

Figure 12.3

The accumulator model for the nonverbal counting process. At each count, the brain increments a quantity, an operation formally equivalent to pouring a cup into a graduated vessel. The final magnitude (the contents of the vessel at the conclusion of the count) is stored in memory, where it represents the numerosity of the counted set. Memory is noisy, which is to say that the values read from memory on different occasions vary. The variability in the values read is proportional to the mean value of the distribution (scalar variability).

(1977) had proposed to explain the generation of mental magnitudes representing durations could be modified to make it generate mental magnitudes representing numerosities. Gibbon had proposed that while a duration was being timed, a stream of impulses fed an accumulator, so that the accumulation grew in proportion to the duration of the stream. When the stream ended (when timing ceased), the resulting accumulation was read into memory, where it represented the duration of the interval. Meck and Church (1983) postulated that to obtain magnitudes representing numerosity, the equivalent of a pulse-former was inserted into the stream of impulses, so that for each count there was a discrete increase in the contents of the accumulator, as happens when a cup of liquid is poured into a graduated vessel (see figure 12.3). At the end of the count, the resulting accumulation is read into memory, where it represents the numerosity.

The model in figure 12.3 is the well-known accumulator model for nonverbal counting by the successive incrementation of mental magnitudes. It is also the origin of the hypothesis that the mental magnitudes representing duration and those representing numerosity are essentially the same, differing only in what it is they refer to. Put another way, both numerosity and duration are represented mentally by real numbers. Meck and Church (1983) compared the psychophysics of number and time representation in the rat and concluded that the coefficient of variation, the ratio between the standard deviation and the mean, was the same, which is further evidence for the hypothesis that the same system of real numbers is used in both cases.

In the course of their work, Meck and Church (1983) were able to estimate the scale factor relating the mental magnitude scales for numerosity and duration. If the mental magnitude, $\hat{n}$, representing numerosity n, is proportional to the numerosity represented, that is, if $\hat{n} = k_1 n$, and if the mental magnitude, $\hat{d}$, representing duration, d, is proportional to the duration represented, that is, if $\hat{d} = k_2 d$, then letting the mental magnitudes be equal (letting $k_1 n = k_2 d$), gives $n(k_2/k_1)d$. The scale factor k_2/k_1 tells us which numerosities are "mentally equivalent" to which durations, where by mentally equivalent we mean "represented by real numbers of the same magnitude." Meck and Church obtained a value of about $0.2\,\mathrm{s}^{-1}$ for k_2/k_1, meaning that each unit increase in the mental magnitude representing numerosity corresponds to the increase generated by prolonging a duration by 0.2 seconds. Thus, an interval 2 seconds in duration generates a mental magnitude that, if it were used to represent a numerosity, would represent a numerosity of 10.

Knowing this scale factor enabled Meck and Church (1983; see also Meck et al. 1985) to do an experiment directly demonstrating that the mental magnitudes representing numerosity and those representing duration were interchangeable. They first taught rats to choose one lever after hearing a noise of 2 seconds' duration and the other lever after hearing a noise of 4 seconds' duration. The rats learned to make this discrimination under partial reinforcement conditions; that is, on 50 percent of the trials even a correct choice did not produce a reward. Because the animal is thereby accustomed to not receiving a reward on many of the trials when it makes a correct choice, this procedure allows the experimenter to give the trained animal unrewarded "probe" trials. In a probe trial, the animal receives a stimulus different from the training stimulus (the reference stimulus) and the question is, how will it judge the probe stimulus? Which reference stimulus will it judge to be "more like" the probe stimulus? Its judgment is indicated by which lever it chooses. In this case, the reference stimuli are represented by the mental magnitudes corresponding to durations of 2 and 4 seconds.

The probe stimuli in the experiment were long sequences of noise bursts. The sequences were much longer than 4 seconds, so the mental magnitudes representing duration would be much greater than either of the reference magnitudes. If the magnitudes representing the duration of a probe sequence were compared with the reference magnitudes, they would be more similar to the greater of those two reference magnitudes, albeit basically outside the range of the reference magnitudes. However, the number of bursts in the sequence ranged from 10 to 20, which is to say that the mental magnitudes produced by the nonverbal counting of these sequences covered the same range of mental magnitudes produced by durations ranging from 2 to 4 seconds. Meck and Church hoped their subjects would make their choice between levers by comparing the mental magnitudes generated by counting the bursts to the reference mental magnitudes, ignoring the fact that the reference magnitudes represented durations while the magnitudes being compared with them on these trials represented numerosity. This was the result they obtained; when there were ten or close to ten noise bursts, rats chose the lever corresponding to the shorter duration; when there were twenty or close to twenty bursts, they chose the lever corresponding to the longer duration. This is strong evidence that a common system of mental magnitudes is used to represent both uncountable (continuous) and countable (discrete) quantities.[2]

Arithmetic Reasoning in Animals

We have repeatedly referred to the real number system because numbers acquire their representational utility by virtue of the fact that they can be arithmetically manipulated: added, subtracted, multiplied, divided, and ordered. From a formal point of view, if mental magnitudes could not be arithmetically manipulated, there would be no justification for calling them numbers. From a formalist perspective, numbers are just entities that are arithmetically manipulable. Thus, when we refer to the real numbers in the brain, we mean magnitudes that can be arithmetically processed in the brain and that refer to countable and uncountable quantities.

There is a considerable experimental literature demonstrating that laboratory animals reason arithmetically with real numbers. They add, subtract, divide, and order subjective durations and subjective numerosities; they divide subjective numerosities by subjective durations to obtain subjective rates of reward; and they multiply subjective rates of reward by the subjective magnitudes of the rewards to obtain subjective incomes. Here we summarize a few of the relevant studies.

Adding Numerosities

Boysen and Berntson (1989) taught chimpanzees to pick the Arabic numeral corresponding to the number of items they observed. In the last of a series of tests of this ability, they had their subjects go around a room and observe either caches of actual oranges in two different locations or simply Arabic numerals that substituted for the caches themselves. When they returned from a trip, the chimps picked the Arabic numeral corresponding to the sum of the two numerosities they had seen, whether the numerosities had been directly observed (hence, possibly counted) or symbolically represented (hence not countable). In the latter case, the magnitudes corresponding to the numerals observed were presumably retrieved from a memory map relating the arbitrary symbols for number (the Arabic numerals) to the mental magnitudes that naturally represent those numbers. Once retrieved, they could be added just like the magnitudes generated by the nonverbal counting of the caches.

Subtracting Durations and Numerosities

In each trial of the time-left procedure (Gibbon and Church 1981), subjects are offered an ongoing choice between a steadily diminishing delay, on the one hand (the time-left option), and a fixed delay, on the other (the standard option). At some unpredictable point in the course of a trial, the opportunity to choose suddenly terminates, and the subject must then endure the delay associated with the option it was exercising at that moment. If it was pecking the standard key, it is stuck with the standard delay; if it was pecking the time-left key, it is stuck with the time left. The initial value of the time left—the value at the beginning of a trial—is much longer than the standard delay, but it grows ever shorter as the trial goes on because the time left is the initial value minus the time so far elapsed in a trial. Therefore, the longer a trial persists before the loss of choice, the better the time-left option becomes relative to the standard. When the subjective time left is less than the subjective standard, the subjects switch from the standard option to the time-left option. The subjective time left is the subjective duration of a remembered initial duration (subjective initial duration) minus the subjective duration of the interval elapsed since the beginning of the trial. Thus, in this experiment the subjects' behavior depends on the subjective ordering of a subjective difference and a subjective standard.

In the number-left procedure (Brannon et al. 2001), pigeons peck a center key to generate flashes and to activate two choice keys. The flashes are generated on a variable ratio schedule, which means that the number of pecks required to generate each flash varies randomly between 1 and 8. When the choice keys are activated, the pigeons can obtain a reward by pecking either of them, but only after their pecks generate the requisite number of flashes. For one of the choice keys, the so-called standard key, the requisite number is fixed and independent of the number of flashes already generated. For the other choice key, the number-left key, the requisite number is the difference between a fixed starting number and the tally of flashes already generated by pecking the center key. The flashes generated by pecking a choice key are also delivered on a variable ratio schedule.

The use of variable ratio schedules for flash generation dissociates time and number. The number of pecks required to generate any given number of flashes—and, hence, the amount of time spent pecking—varies greatly from trial to trial. This makes possible an analysis to determine whether the subjects' choices are controlled by the time spent pecking the center key or by the number of flashes thus generated.

In this experiment, the subjects chose the number-left key when the subjective number left was less than some fraction of the subjective number of flashes required on the standard key. Thus, their behavior was controlled by the subjective ordering of a subjective numerical difference and a subjective numerical standard.

There is also evidence that the mental magnitudes representing duration and rates are signed; that is, there are both positive and negative mental magnitudes (Gallistel and Gibbon 2000; Savastano and Miller 1998). In other words, there is evidence not only for subtraction but for the hypothesis that the system for arithmetic reasoning with mental magnitudes is closed under subtraction.

Dividing Number by Duration

When vertebrates from fish to humans are free to forage in two different nearby locations, moving back and forth repeatedly between them, the ratio of the expected durations of the stays in the two locations matches the ratios of the numbers of rewards obtained per unit of time (Herrnstein 1961). Until recently, it had been assumed that this "matching" behavior depended on the law of effect. When subjects do not match, they get more reward per unit of time invested in one patch than per unit of time invested in the other. Only when they match do they get equal returns on their investment. Thus, matching could be explained on the assumption that subjects try different ratios of investments (different ratios of expected stay durations) until they discover the ratio that equates the returns (Herrnstein and Vaughan 1980).

Despite the plausibility of this explanation, it has never been possible to construct a model based on this assumption that predicted the details of the behavior at all well (Lea and Dow 1984). For one thing, because of the way rewards are scheduled in the customary experimental paradigm, the return to be expected from a given location increases the longer that location has gone unvisited. If the subject's decisions to leave one location to sample the other were sensitive to the returns on its behavioral investments, then the probability of leaving ought to get higher as the duration of a stay increases, but it does not. The probability of leaving is Markovian (Heyman 1979); that is, it looks statistically as if the subjects repeatedly flipped a coin to decide whether to leave (Gibbon 1995; Gibbon et al. 1988), with the outcome of later flips being independent of the outcome of earlier flips. This discovery led to the suggestion that matching behavior is an unconditioned response to the experience of a given ratio of rates of reward (Heyman 1982).

Recently, Gallistel et al. (2001) have shown that rats adjust to changes in the scheduled rates of reward as fast as it is in principle possible to do so; they are "ideal detectors" of such changes. They could not adjust anywhere near as rapidly as they in fact do adjust if they were discovering by trial and error the ratio of expected stay durations that equated their returns. This means that Heyman (1982) was right; matching behavior is an unconditioned or preprogrammed response to the experience of different rates of reward. The importance of this in the present context is that a rate is a countable quantity—the number of rewards received in a given interval—divided by an uncountable quantity—the duration of the given interval.

Gallistel and Gibbon (2000) review the evidence that both Pavlovian and instrumental conditioning depend on subjects' estimating rates of reward. They argue that rate of reward is the fundamental variable in conditioned behavior. The importance of this in the present context is twofold. First, it is evidence that subjects divide mental magnitudes. Second, it shows why it is essential that countable and uncountable quantities be represented by commensurable mental symbols, symbols that are all part of the same system and can be arithmetically combined without regard to whether they represent countable or uncountable quantities. If countable quantities were represented by one system (say, a system of discretely ordered symbols) and uncountable quantities by a different system (a system of continuously ordered magnitudes), it would not be possible to estimate rates.

Multiplying Rate by Magnitude

When the magnitudes of the rewards obtained in two different locations differ, then the ratio of the expected stay durations is determined by the ratio of the incomes obtained from the two locations (Catania 1963; Harper 1982; Keller and Gollub 1977; Leon and Gallistel 1998). The income from a location is the product of the rate and the magnitude. Thus, this result implies that subjects multiply subjective rate by subjective magnitudes to obtain subjective incomes. The signature of multiplicative combination is that changing one variable by a given factor—for example, doubling the rate—changes the product by the same factor (doubles the income) regardless of the value of the other factor (the magnitude of the rewards). Leon and Gallistel (1998) showed that changing the ratio of the rates of reward by a given factor changed the ratio of the expected stay durations by that factor, regardless of the ratio of the reward magnitudes, thereby proving that subjective magnitudes combine multiplicatively with subjective rates to determine the ratio of expected stay durations.

Ordering Numerosities

Most of the paradigms that demonstrate mental addition, subtraction, multiplication, and division also demonstrate the mental ordering of the mental magnitudes because the subject's choice depends on it. Brannon and Terrace (2000) demonstrated more directly that monkeys order numerosities by simultaneously presenting several arrays differing in the numerosity of the items constituting each array and requiring their macaque subjects to touch the arrays in the order of their numerosity. When the subjects had learned to do this for numerosities between 1 and 4, they generalized immediately to numerosities between 5 and 9. Perhaps most important, it was impossible to teach the subjects to touch the arrays in an order that did not conform to the order of the numerosities. This implies that the ordering of the numerosities is highly salient for a monkey a priori (before training).

In summary, research with vertebrates, some of which have not shared a common ancestor with man since before the rise of the dinosaurs, implies that they represent both countable and uncountable quantities by mental magnitudes (real numbers). The system of arithmetic reasoning with these mental magnitudes is closed under the basic operations of arithmetic; that is, mental magnitudes may be mentally added, subtracted, multiplied, divided, and ordered without restriction.

Evidence that Humans Represent Numerosity with Mental Magnitudes

Symbolic Size and Distance Effects

From an evolutionary standpoint, it would be odd if humans did not share with their remote vertebrate cousins (pigeons) and near vertebrate cousins (macaques and chimpanzees) the mental machinery for representing countable and uncountable quantities by a system of real numbers. That humans do represent numbers with mental magnitudes was suggested by Moyer and Landauer (1967, 1973) when they discovered what have come to be called symbolic size and distance effects. When subjects are asked to judge the numerical order of Arabic numerals as rapidly as possible, their reaction time is determined by the relative numerical distance. The greater the distance between the two numbers, the more quickly their order may be judged (the distance effect), and for a fixed difference, the greater the magnitude of the two numbers, the longer it takes to judge their order (the size effect).

Moyer and Landauer suggested that the effects of numerical magnitude on reaction time implied that Weber's law applied to symbolically represented numerical magnitudes. Weber's law is that the discriminability of two quantities is a function of their ratio. Moyer and Landauer (1973) suggested that symbolically represented numbers were translated into mental magnitudes in order to judge the numerical ordering of the represented numbers, and that noise in the mental magnitudes made it more difficult to determine which magnitude was greater. This, together with the assumption that more difficult discriminations take longer to make, explains the symbolic distance effect.

Weber's law is often taken to imply logarithmic compression in the mapping between objective and subjective magnitudes. This assumption, together with the usually implicit (and physically implausible) assumption that the noise in mental magnitudes is magnitude independent, gives Weber's law because when mapped onto a logarithmic scale, magnitudes with a given ratio are separated by a given distance.

The assumption of logarithmic compression is, however, inconsistent with the results of the time-left and number-left experiments discussed earlier. If mental magnitudes were proportional to the logarithms of objective magnitudes, then equal differences in mental magnitudes would correspond to equal ratios between the corresponding objective magnitudes. Thus, when subjects are asked to compare the subjective difference between two numbers against some standard, the results should depend, not on the objective difference between the two magnitudes, but rather on their ratio. If the subjective difference between 6 and 3 is equal to the subjective magnitude of 3, then the subjective difference between 60 and 30 should likewise be equal to the subjective magnitude of 3, because $\log(60) - \log(30) = \log(6) - \log(3) = \log(2)$. This is implausible a priori, and it is contrary to experimental fact. In both the time-left and number-left experiments, the point of subjective equality increased linearly with the (objective) difference between the initial and standard magnitudes when their ratio was held constant.

The just-mentioned experimental results imply that the mental magnitudes representing both numerosity and duration are approximately scalar mappings of the objective magnitudes. If these mental magnitudes have scalar noise, then this too gives Weber's law; the discriminability of two such magnitudes will depend on their ratio, not their difference. Thus, the symbolic size and distance effects are consistent with the assumption that when humans judge numerical order, they represent number by mental magnitudes with scalar variability, just as do other vertebrates. What is unique in humans (and a few chimpanzees trained by humans) is that these mental magnitudes can be evoked by a learned mapping from the culturally defined linguistic and graphemic symbols for the integers to the mental magnitudes that represent numerosity in the nonverbal or preverbal brain.

Nonverbal Counting in Humans

Given the evidence from the symbolic distance effect that humans represent number with mental magnitudes, it seems likely that they share with the nonverbal animals in the vertebrate clade a nonverbal counting mechanism that maps from numerosities to the mental magnitudes that represent them. If so, then it should be possible to demonstrate nonverbal counting in humans when verbal counting is suppressed. Whalen et al. (1999) presented subjects with Arabic numerals on a computer screen and asked them to press a key as fast as they could without counting until it felt like they had pressed the number signified by the numeral. The results from humans looked very much like the results from pigeons and rats (figure 12.4). The mean number of presses increased in proportion to the target number, and the standard deviations of the distributions of presses increased in proportion to their mean, so that the coefficient of variation was constant.

This result suggests first, that the subjects could count nonverbally, and second, that they could compare the mental magnitude thus generated with a magnitude obtained by the learned mapping from numerals to mental magnitudes. Finally, it implies that

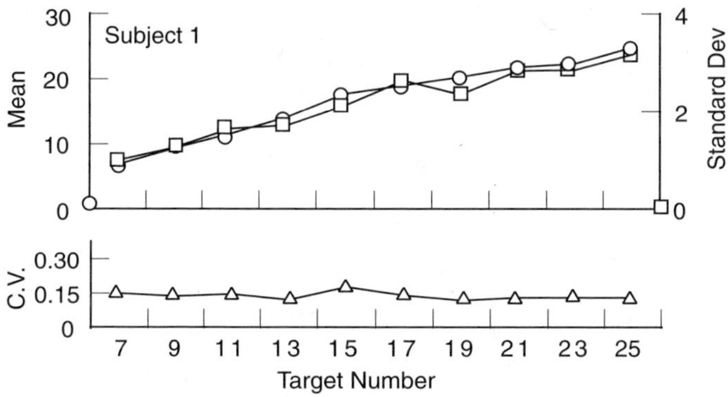

Figure 12.4
Representative data from the human nonverbal counting experiment by Whalen et al. (1999). The mean number of presses made increased in proportion to the target number (top panel, left ordinate) and so did the variability (top panel, right ordinate), so the coefficient of variation was constant (bottom panel). Human nonverbal counting exhibits the same scalar variability as nonhuman counting and timing (compare with figures 12.1 and 12.2).·

the mapping from numerals to mental magnitudes is such that the mental magnitude given by this mapping approximates the mental magnitude generated by counting the numerosity signified by a given numeral: nonverbally counting a three-item set and seeing the numeral "3" both yield approximately the same mental magnitude.

In a second task, the subjects observed a dot flashing rapidly but at irregular intervals. The rate of flashing (eight per second) was about twice as fast as estimates of the maximum speed of verbal counting (Mandler and Shebo 1982). The subjects were asked not to count, but to say about how many times they thought the dot had flashed. As in the first experiment, the mean number estimated increased in proportion to the number of flashes, and the standard deviation of the estimates increased in proportion to the mean estimate. This implies that the mapping between the mental magnitudes generated by nonverbal counting and the verbal symbols for numerosities is bidirectional. It can go from a symbol to a mental magnitude that is comparable to the one that would be generated by nonverbal counting, and it can go from the mental magnitude generated by a nonverbal count to a roughly corresponding verbal symbol. In both cases, the variability in the mapping is scalar.

Whalen et al. (1999) gave several reasons for believing that their subjects did not count subvocally. We will not review them here because recent further experiments by Cordes et al. (2002) speak more directly to this issue. Cordes et al. suppressed articulation by having their subjects repeat "the" coincident with each press while they attempted to press a button a target number of times. This manipulation certainly suppresses audible articulation. Cordes et al. recorded the subjects while they pressed, and there was no audible counting. This manipulation presumably suppresses subvocal articulation as well, because it does not seem likely that someone can subvocally articulate one word (a counting word) at the same moment they audibly articulate a different, noncounting word. The manipulation of suppressing articulatory coding—saying "the" with every press—was particularly effective in that the subjects found it easy to do and it required them to articulate a noncounting word at the very moment when they would articulate a counting word if they were verbally counting.

In control experiments, the subjects were asked to count their presses out loud in one of two ways: the conventional way, fully pronouncing each counting word, and a way that they found much easier, which was to use only the single-digit counting words, silently keeping track of the tens count. In all conditions, the subjects were asked to press as fast as possible.

The variability data from the condition in which subjects were required to say "the" coincident with each press are shown in figure 12.5 (filled squares). As in Whalen et al. (1999), the coefficient of variation was constant (scalar variability). The best-fitting line has a slope that does not differ significantly from 0. The contrasting results from the control conditions, where the subjects counted out loud, fully pronouncing each counting word, are the open squares. Here the slope on this log-log plot deviates significantly from 0. In verbal counting, one would expect counting errors—double counts and skips—to be the most common source of variability. On the assumption that the probability of a counting error is approximately the same at successive steps in a count, the resulting variability in final counts should be binomial rather than scalar. It should increase in proportion to the square root of the target value, rather than in proportion to the target value. If the variability is binomial rather than scalar, then when the coefficient of variation is plotted against the target number on a log-log plot, it should form a straight line with a slope of −0.5. This is what was in fact observed in the out-loud counting conditions. The variability was much less than in the nonverbal counting conditions and, more important, it was binomial rather than scalar. The mean slope of the subject-by-subject regression lines in the two control conditions was significantly less than 0 and was not significantly different from −0.5. The contrasting patterns of variability in the counting out-loud and nonverbal counting conditions considerably strengthen the evidence against the hypothesis that the subjects in the nonverbal counting conditions were counting subvocally.

A second feature of the data from some subjects in the Cordes et al. experiment further strengthens the evidence against the subvocal counting hypothesis. In two of the subjects, the relation between the target number and the mean number of presses was a power function with an exponent significantly greater than 1, that is, with a slight upward curve (figure 12.6). Power functions describe the relation between objec-

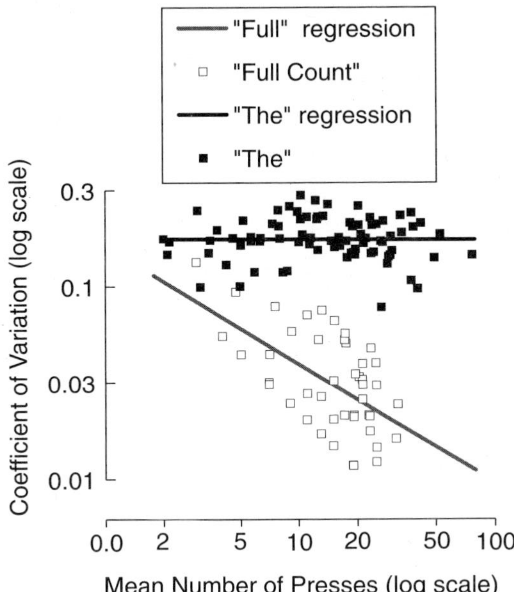

Figure 12.5
The coefficients of variation (σ/μ) are plotted against the numbers of presses for the conditions in which subjects counted nonverbally and for the condition in which they fully pronounced each counting word (double logarithmic coordinates). In the former condition, there is scalar variability, that is, a constant coefficient of variation. The slope of the regression line relating the log of the coefficient of variation to the log of the mean number of presses does not differ from 0. In the latter, the variability is much less and it is binomial; the coefficient of variation decreases in proportion to the square root of the target number. In the latter case, the slope of the regression line relating the log of the coefficient of variation to the log of the mean number of presses differs significantly from 0 but does not differ significantly from –0.5, which is the slope predicted by the binomial variability hypothesis. Data from Cordes et al. (2001).

tive and subjective magnitudes (Stevens 1970). Their most interesting property is that they preserve equal proportions. Any two pairs of objective magnitudes that have the same objective ratio (for example, 3 : 2 and 30 : 20) have the same subjective ratio when the mapping from objective to subjective magnitudes is a power function.

This finding is consistent with the scalar version of the magnitude mapping hypothesis by which the symbolic distance effect is generally explained—the hypothesis that number words and numerals map to the same kind of mental magnitudes that represent continuous quantities like stimulus intensities, and that these remembered mental magnitudes have scalar noise. It suggests that the mapping is constructed by arranging the linguistic symbols along one continuum and the numerical magnitudes

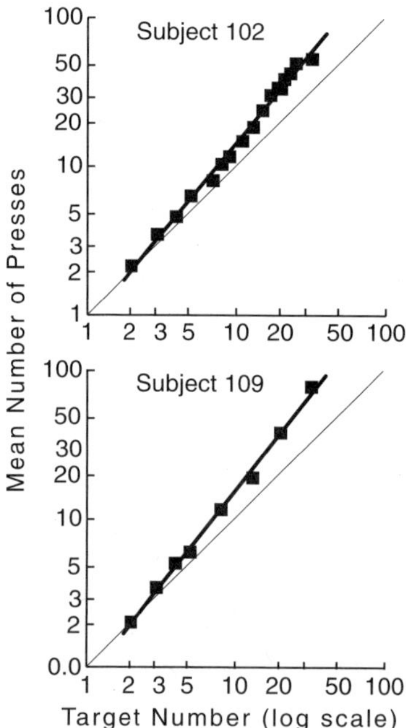

Figure 12.6

The mean number of presses is plotted against the target number for two of the four subjects in Cordes et al. who responded for targets ranging from 2 to 32 (double logarithmic coordinates). For these two subjects, the points fall on straight regression lines (thick lines) whose slopes are slightly but significantly greater than 1 (thin lines). Note that this deviation from a scalar relation (a line with a slope of 1) is evident even in the small number range (2 to 5). Data from Cordes et al. (2001).

along an orthogonal continuum, as in a conventional scatter graph. The loci of points relating the two continua (the points that define the mapping, that is, the function relating the two continua) fall on a power curve whose slope is close to but not necessarily equal to 1. Learning the meaning of the counting words, in this hypothesis, involves learning the appropriate parameters for the power function that relates the continuum on which the symbols are arranged to the continuum that represents numerosity.

On the other hand, it is hard to understand how subvocal counting could yield a power function relation between the target count and the mean number of presses made that has a slope different from 1. Presumably with vocally based counting, the subject articulates each successive counting word coincident with each successive press until the articulated word matches the articulation of the target number. In their haste, some subjects might skip (fail to count) some percentage of the presses they make, so that they systematically undercount their presses, but this would lead to scalar error, not an error in the exponent. The plot of the log of the mean number of presses against the log of the target number should still be a straight line with slope 1, because any scalar relation looks like this on a log-log plot.

In sum, nonverbal counting may be demonstrated in humans, and it looks just like nonverbal counting in nonhumans. Moreover, mental magnitudes (real numbers) comparable to those generated by nonverbal counting appear to mediate judgments about the numerical ordering of symbolically presented integers. This suggests that the nonverbal counting system is what underlies and gives meaning to the linguistic representation of numerosity.

Where the Integers Come From

In our hypothesis, it is the real numbers, not the integers, that are the primitive foundation of numerical reasoning. The integers are a special case whose prominence in the cultural history of numbers derives from the discrete character of language. When a discrete system like language attempts to represent a quantity, it will find it much easier to represent a countable (discrete) quantity than an uncountable (continuous) quantity.

Gallistel and Gelman (1992) argue that the nonverbal domain of numerical estimation and arithmetic reasoning in animals is operative in the very young child and provides the foundation on which the child's understanding of verbally mediated arithmetic estimation and reasoning is based. The nonverbal mechanism for establishing reference is the accumulator counting mechanism, which produces mental magnitudes that represent numerosities. By "represent," we mean that the symbols generated (the mental magnitudes) both refer to numerosities and enter into arithmetic reasoning operations.

The situations that trigger verbal counting also trigger nonverbal counting. It is important to note that the verbal counting process is homomorphic to the nonverbal counting process. In particular, both processes have effective procedures for defining successor symbols. Each step in the verbal process summons the next word from the list of counting words. Each count in the nonverbal process defines a next magnitude. Thus the products of both processes are discretely ordered; that is, the magnitude that results from the next step is always one fixed increment greater than the magnitude that results from the previous step, and the word used in the next verbal count is always one item later in the list than the preceding word. Finally, the product of the verbal process as a whole, that is, the word used in the last step, represents the numerosity of the set being counted, as does the magnitude produced by the final increment in the nonverbal counting process.

Gallistel and Gelman (1992) argue that the child perceives the homomorphism between the nonverbal and the verbal counting process, and this leads to the assumption that the words used in the counting process represent the same aspect of the world as do the mental magnitudes obtained from the nonverbal counting process. This means two things: First, it means that the child thinks that the counting words refer to the same things in the world as the mental magnitudes representing numerosity refer to, namely, countable quantities. Second, it means that the rules of inference (the rules of operation) governing the magnitudes will be the rules of inference that govern the use of numbering words. This is the structural mapping (homomorphism) between the verbal and the nonverbal systems.

In short, we suggest that the integers are picked out by language because they are the magnitudes that represent countable quantities. A countable quantity is the only kind of quantity that can readily be represented by a system founded on discrete symbols, as language is. It is language that makes us think that God made the integers, because the learning of the integers is the beginning of linguistically mediated mathematical thinking about both countable and uncountable quantities.

The Problem of Exact Equivalence

This hypothesis raises interesting questions about where some of our intuitive convictions about quantity come from. One of these concerns our concept of exact equivalence. Empirically, there is no such thing as exact equivalence among uncountable quantities, as every dressmaker knows. Two measured quantities are never exactly the same. If we relied simply on our experience of uncountable quantities, we would not have a concept of exact equivalence, because no two experienced lengths and no two experienced durations are ever exactly the same.

Nonetheless, we believe that when equals are added to equals, the results are equal. We believe this despite the fact that it may or may not be true of mental magnitudes, depending upon whether the nonverbal system for reasoning with magnitudes recognizes equivalence (substitutability) and how it decides whether two magnitudes are equivalent (substitutable). On the face of it, the mental magnitude generated by adding a unit of magnitude (the counting increment) to the remembered magnitude corresponding to "5" on one occasion will not be exactly the same mental magnitude obtained by adding a unit of magnitude to this "same" remembered magnitude on another occasion because the remembered magnitude itself varies from occasion to occasion. This realization has been an obstacle to the more general acceptance of the hypothesis that the real numbers are a psychologically primitive system for representing both uncountable and countable quantities (Carey 1998, 2001; Hauser and Carey 1998; Leslie et al. 1998).

At least three answers to the problem of equivalence suggest themselves. First, a notion of quantitative equivalence based on real numbers must be essentially a statistical notion. Two noisy mental magnitudes must be judged to represent equivalent quantities only if they are not decideably different, that is, if they are not reliably orderable. A system that works with real numbers cannot determine equivalence by exact comparison because, given the noise in the representational system itself, no two mental magnitudes ever match exactly. Two noisy magnitudes can, however, be so close that repeated attempts to determine their ordering lead to the conclusion that they cannot be reliably ordered. This can be taken as equivalent to the conclusion that they are substitutable one for the other. In other words, if a and b are magnitudes, then $a = b$ just in case neither $a > b$ nor $b > a$, or, just in case $a \geq b$ and $b \geq a$. In this account, we believe that when equals are added to equals, the results are equal because it is a truth about our nonverbal processing of mental magnitudes, whose processing of order and equivalence is adapted to the noisiness of the symbols being processed.

A slightly different answer that suggests itself is that the preverbal system of arithmetic reasoning makes use of computational shortcuts that have implicit in them principles about the outcomes of arithmetic processing. A system that reasoned arithmetically with noisy magnitudes might implicitly assume—rather than empirically test for—the equivalence of adding a unit of magnitude to equivalent remembered magnitudes. When dealing cards, for example, it might not compute through each round the numbers of cards the players have and whether those numbers are equal. It might only take care to deal one and only one card to every player on each round. If it operates as if this care guaranteed numerical equality, then it operates in accord with the principle that when equals are added to equals the results are equal. This principle eliminates the need to actually do the sums and compare them at the end of every round. The system does not have to keep four running sums because it knows that the sums remain equal as long as it continues to add equal increments to each on every round.

As this example shows, implicit principles about the outcomes of arithmetic procedures could produce significant computational and mensurational economies. They may also guide verbally mediated reasoning about quantity.

A third answer that suggests itself is that the discrete nature of the verbal representation of countable quantity is the origin of our notion of exact equivalence. Words are discrete entities; the word "three" is not confusable with an infinite number of other words that are not really "three" but that are arbitrarily "close" to "three."

In fact, it is unclear what "close" could mean when it comes to words. Thus, the outcomes of two carefully done verbal counts of the same set will yield the same counting word to represent the set, as every bank teller knows. As already suggested, the outcomes of two different nonverbal counts of the same set will not contradict this because their ordering will not be reliably decidable. Thus, language might not only pick out the integers, it may also highlight a discrete notion of exact equivalence. In the absence of language, exact equivalence may not be an issue. It is not easy to think of a nonalgebraic context in which exact equivalence is of any consequence.

In any case, we do not believe that uncertainty about where our notion of exact equivalence comes from should blind us to the experimental evidence that human numerical reasoning uses noisy magnitudes to determine such fundamental things as numerical order, even when it is given linguistic symbols that represent very small numerosities. If our underlying nonverbal symbols for twoness, threeness, and fourness are fundamentally discrete, and if the "4," "3," and "2" acquire their meaning by reference to these discrete nonverbal symbols (cf. Carey 2001), then it is hard to see why it takes us longer to decide that fourness is larger than threeness than it does to decide that fourness is larger than twoness.

Is There a Discrete Foundation for Integers in the Perception of Small Numerosities?
It is widely accepted that the symbolic size and distance effects imply that adult humans map linguistic symbols for number to mental magnitudes and that they rely on the comparison of those noisy magnitudes to determine numerical order. At least at present, there is no other explanation for these experimentally well-established effects. As already explained, these effects imply that our underlying representation of numerosity has the continuous character of real numbers rather than the discrete character of integers. Nonetheless, it is commonly argued that our concept of an integer originates in a preverbal system for representing numerosity that itself uses countable (discrete) rather than uncountable (continuous) sets of symbols (Carey 2001; Leslie et al. 1998; Simon 1999). It is generally assumed that this discrete system represents only small numerosities (four or less) and that the mapping from small numerosities to the discrete mental entities that represent them—the so-called subitizing process—does not employ any form of counting. In this hypothesis, the numerosity of small sets—oneness, twoness, threeness, and fourness—is directly perceived—like orangeness, cowness, treeness, and forkness. Alternatively, it has been suggested that the numerosity of small sets is implicitly represented by the numerosity of the set of object files that the perceptual system opens, but that numerosity is not explicitly symbolized (Carey 2001).

We see several empirical and theoretical problems with this hypothesis. The existence of a subitizing process for the direct and unvarying apprehension of small numerosities has often been argued on the basis of empirically shaky claims about the form of the reaction time function for rapid numerical estimation (for example, Davis and Pérusse 1988; Siegler and Robinson 1982; Trick and Pylyshyn 1994). It has been claimed either that this function is flat for numerosities between 1 and 4 or 5, or that there is a discontinuity in the slope of this function somewhere at 4 or 5. Neither claim is consonant with the results from several careful determinations of this function (Balakrishnan and Ashby 1992; Folk et al. 1988), which show that the reaction time to judge a numerosity is longer the greater the numerosity, and that each increment in reaction time is greater than the preceding increment (that is, the function accelerates over the range from 1 to 4). The reaction time function for adult judgments of numerosity is at least as consistent with a counting model as it is with a direct perception model (Gallistel and Gelman 1991). If fourness is like forkness and twoness like cowness, as some versions of the discrete-origins hypothesis maintain, then one needs to explain why it takes so much longer to perceive fourness than it does to perceive twoness.

Second, if small numerosities were represented differently from large numerosities, then one would expect to see a discontinuity in the psychophysically measurable properties of number representations at the point where one form of representation gives way to the other. The data from the experiments by Cordes et al. (2001) in which adults counted key presses while saying "the" coincident with each press are relevant here. For four of the subjects, the target numbers included 2, 3, 4, and 5, as well as several larger numbers that were unequivocally beyond the range of the putative subitizing process. The coefficient of variation was the same for these small numbers as for the large numbers (see figure 12.5, filled squares). Moreover, in the two subjects whose mapping from linguistically represented numerosity to mental magnitudes was systematically distorted, this distortion was continuous between the small and large number ranges (figure 12.6).

Proving continuity experimentally is like proving the null hypothesis; it cannot be done. However, there is no evidence of discontinuity in the psychophysical evidence from adult humans, and this is hard to reconcile with the hypothesis that small and large numbers are represented in fundamentally different ways.

We think the theoretical problems with the hypothesis are at least as great as the empirical problems. If there are discrete nonverbal representations of numerosity for only the small numbers—either nonverbal symbols or percepts for oneness, twoness, threeness, and fourness, or implicit representations by sets of mental entities whose numerosity equals the objective numerosity—then the first question that arises is whether there is any arithmetic processing of these symbols (or sets of object files). Can the percept of "oneness" be mentally added to the percept of "twoness" to get the percept of "threeness"? Can one set of object files be compared with another set of object files to determine if the two sets have the same numerosity? If so, then to make the comparison of two sets of three each, the mind would have to have six object files in play at once, and this is, as we understand it, thought to be impossible.

Can a set of one object file be combined with a set of two object files to yield a set of three object files? If so, then the assumption that these discrete symbols exist for only a few small numbers leads immediately to problems with closure. This system will not be closed even under addition, because there will be no symbol to represent the results of adding "threeness to threeness." Thus, if operations with these very limited sets of mental symbols are the foundation of numerical understanding, it is a puzzle how we come to believe in the infinite extensability of number, in the fact that you can always add one more (Hartnett and Gelman 1998). On the other hand, if these symbols for numerosity—whether explicit or implicit—cannot be processed arithmetically, then what justification is there for saying that these symbols constitute a numerical representation? If a representation does not enable any of the processes appropriate for numbers, then one must ask why it may be said to be a representation of number.

The second theoretical problem concerns the relation between the two forms of numerical representation, one discrete and one continuous. The two systems would seem to be immiscible for the same reasons that analog and digital computers cannot be hybridized. Although both do arithmetic, they do it in fundamentally different ways. Thus, there is no way of adding a digitally represented magnitude (for example, a bit pattern) to a magnitude represented by an analogical magnitude (for example, a voltage) because the two forms of representation are immiscible. It is hard to see why this same problem does not arise in the developing human mind, if it represents some numbers discretely and others by means of magnitudes. If oneness is represented discretely but tenness is represented by a mental magnitude, how is it possible to mentally add oneness to tenness?

This question—the question of where the human conception of an integer comes from—is currently one of the most controversial in the field of numerical cognition (Carey 2001; Leslie et al. 1998; Simon 1999). Clearly, our hypothesis about the relation between mental magnitudes and the linguistically mediated concept of a number cannot be more widely embraced until consensus is reached on this central question.

Summary

The evidence from experiments that probe the properties of numerical representations in nonverbal animals and humans suggests that there exists a common system for representing both countable and uncountable quantities by mental magnitudes that are formally equivalent to real numbers. These mental magnitudes are arithmetically processed without regard to whether they represent countable or uncountable quantities.

Adult humans appear to rely on a mapping from the linguistic symbols for number to these preverbal mental magnitudes, even for answering elementary verbal or written questions like, "is 3 greater than 2?". This has led us to suggest that the nonverbal system for arithmetic reasoning with mental magnitudes precedes the verbal system both phylogenetically, and ontogenetically, and that the verbal symbols for numerosity are given their meaning by reference to the nonverbal mental magnitudes that represent countable quantity. If this suggestion is correct, then the real numbers are the psychologically primitive system, not the natural numbers. The special role of the natural numbers in the cultural history of arithmetic is a consequence of the discrete character of human language, which picks out of the system of real numbers in the brain the discretely ordered subsets generated by the nonverbal counting process, and makes these the foundation of the linguistically mediated conception of number.

Notes

1. This review was completed in 2001.

2. After completion of this review, Balci and Gallistel (2005, submitted) replicated this experiment with human subjects. However, the results of further controls led them to a different interpretation: They suggest that transfer from duration judgment to number judgment is based on comparable proportions between individual probe stimuli, whether duration or number, and the tested range of stimuli of that kind. When confronted with a number probe subjects judge it on the basis of where it falls within the range of number probes tested. And likewise when confronted with a duration probe. Comparable positions within the respective ranges yield comparable judgments (long/short, large/small). In this interpretation, no comparison between mental magnitudes representing different kinds of quantities (number and duration) is involved—only comparison of (unit less) proportions.

References

Balakrishnan, J. D., and Ashby, F. G. 1992. Subitizing: Magical numbers or mere superstition. *Psychological Research* 54: 80–90.

Balci, F., and Gallistel, C. R. (submitted). Spontaneous transfer from duration discrimination to number discrimination: Is it all a matter of proportion?

Boysen, S. T., and Berntson, G. G. 1989. Numerical competence in a chimpanzee (*Pan troglodytes*). *Journal of Comparative Psychology* 103: 23–31.

Brannon, E. M., and Terrace, H. S. 2000. Representation of the numerosities 1–9 by rhesus macaques (*Macaca mulatta*). *Journal of Experimental Psychology: Animal Behavior Processes* 26(1): 31–49.

Brannon, E. M., Wusthoff, C. J., Gallistel, C. R., and Gibbon, J. 2001. Numerical subtraction in the pigeon: Evidence for a linear subjective numeral scale. *Psychological Science* 12: 238–243.

Carey, S. 1998. Knowledge of number: Its evolution and ontogeny. *Science* 282: 641–642.

Carey, S. 2001. Cognitive foundations of arithmetic: Evolution and ontogenisis. *Mind and Language* 16(1): 37–55.

Catania, A. C. 1963. Concurrent performances: A baseline for the study of reinforcement magnitude. *Journal of the Experimental Analysis of Behavior* 6: 299–300.

Cordes, S., Gelman, R., Gallistel, C. R., and Whalen, D. 2001. Variability signatures distinguish verbal from nonverbal counting for both large and small numbers. *Psychonomic Bulletin and Review* 8: 698–707.

Davis, H., and Pérusse, R. 1988. Numerical competence in animals: Definitional issues, current evidence, and a new research agenda. *Behavioral and Brain Sciences* 11: 561–615.

Dehaene, S. 1997. *The number sense.* Oxford: Oxford University Press.

Folk, C. L., Egeth, H., and Kwak, H. 1988. Subitizing: Direct apprehension or serial processing? *Perception and Psychophysics* 44(4): 313–320.

Gallistel, C. R. 1990. *The organization of learning.* Cambridge, Mass.: Bradford Books/MIT Press.

Gallistel, C. R. 1999. Can a decay process explain the timing of conditioned responses? *Journal of the Experimental Analysis of Behavior* 71: 264–271.

Gallistel, C. R., and Gelman, R. 1991. Subitizing: The preverbal counting process. In W. Kessen, A. Ortony, and F. Craik (eds.), *Memories, thoughts and emotions: Essays in honor of George Mandler* (pp. 65–81). Hillsdale, N.J.: Erlbaum.

Gallistel, C. R., and Gelman, R. 1992. Preverbal and verbal counting and computation. *Cognition* 44: 43–74.

Gallistel, C. R., and Gelman, R. 2000. Non-verbal numerical cognition: From reals to integers. *Trends in Cognitive Sciences* 4: 59–65.

Gallistel, C. R., and Gibbon, J. 2000. Time, rate and conditioning. *Psychological Review* 107: 289–344.

Gallistel, C. R., Mark, T. A., King, A. P., and Latham, P. E. (2001). The rat approximates an ideal detector of changes in rates of reward: Implications for the law of effect. *Journal of Experimental Psychology: Animal Behavior Processes*, 27: 354–372.

Gibbon, J. 1977. Scalar expectancy theory and Weber's law in animal timing. *Psychological Review* 84: 279–335.

Gibbon, J. 1995. Dynamics of time matching: Arousal makes better seem worse. *Psychonomic Bulletin and Review* 2(2): 208–215.

Gibbon, J., and Church, R. M. 1981. Time left: Linear versus logarithmic subjective time. *Journal of Experimental Psychology: Animal Behavior Processes* 7(2): 87–107.

Gibbon, J., Church, R. M., Fairhurst, S., and Kacelnik, A. 1988. Scalar expectancy theory and choice between delayed rewards. *Psychological Review* 95: 102–114.

Harper, D. G. C. 1982. Competitive foraging in mallards: Ideal free ducks. *Animal Behavior* 30: 575–584.

Hartnett, P., and Gelman, R. 1998. Early understandings of numbers: Paths or barriers to the construction of new understandings? *Learning and Instruction* 8(4): 341–374.

Hauser, M., and Carey, S. 1998. Building a cognitive creature from a set of primitives: Evolutionary and developmental insights. In C. Allen and D. Cummings (eds.), *The evolution of mind* (pp. 51–106). Oxford: Oxford University Press.

Herrnstein, R. J. 1961. Relative and absolute strength of response as a function of frequency of reinforcement. *Journal of the Experimental Analysis of Behavior* 4: 267–272.

Herrnstein, R. J., and Vaughan, W. J. 1980. Melioration and behavioral allocation. In J. E. R. Staddon (ed.), *Limits to action: The allocation of individual behavior* (pp. 143–176). New York: Academic Press.

Heyman, G. M. 1979. A Markov model description of changeover probabilities on concurrent variable-interval schedules. *Journal of the Experimental Analysis of Behavior* 31: 41–51.

Heyman, G. M. 1982. Is time allocation unconditioned behavior? In M. Commons, R. Herrnstein, and H. Rachlin (eds.), *Quantitative analyses of behavior*, Vol. 2: *Matching and maximizing accounts* (pp. 459–490). Cambridge, Mass.: Ballinger Press.

Keller, J. V., and Gollub, L. R. 1977. Duration and rate of reinforcement as determinants of concurrent responding. *Journal of the Experimental Analysis of Behavior* 28: 145–153.

Lea, S. E. G. and Dow, S. M. 1984. The integration of reinforcements over time. In J. Gibbon and L. Allan (eds.), *Timing and time perception*. Vol. 423 (pp. 269–277). New York: Annals of the New York Academy of Sciences.

Leon, M. I., and Gallistel, C. R. 1998. Self-stimulating rats combine subjective reward magnitude and subjective reward rate multiplicatively. *Journal of Experimental Psychology: Animal Behavior Processes* 24(3): 265–277.

Leslie, A. M., Xu, F., Tremoulet, P. D., and Scholl, B. 1998. Indexing and the object concept: Developing what and where systems. *Trends in Cognitive Sciences* 2: 10–18.

Mandler, G., and Shebo, B. J. 1982. Subitizing: An analysis of its component processes. *Journal of Experimental Psychology: General* 11: 1–22.

Mechner, F. 1958. Probability relations within response sequences under ratio reinforcement. *Journal of the Experimental Analysis of Behavior* 1: 109–122.

Meck, W. H., and Church, R. M. 1983. A mode control model of counting and timing processes. *Journal of Experimental Psychology: Animal Behavior Processes* 9: 320–334.

Meck, W. H., Church, R. M., and Gibbon, J. 1985. Temporal integration in duration and number discrimination. *Journal of Experimental Psychology: Animal Behavior Processes* 11: 591–597.

Moyer, R. S., and Landauer, T. K. 1967. Time required for judgments of numerical inequality. *Nature* 215: 1519–1520.

Moyer, R. S., and Landauer, T. K. 1973. Determinants of reaction time for digit inequality judgments. *Bulletin of the Psychonomical Society* 1: 167–168.

Platt, J. R., and Johnson, D. M. 1971. Localization of position within a homogeneous behavior chain: Effects of error contingencies. *Learning and Motivation* 2: 386–414.

Roberts, W. A., Coughlin, R., and Roberts, S. 2000. Pigeons flexibly time or count on cue. *Psychological Science* 11(3): 218–222.

Savastano, H. I., and Miller, R. R. 1998. Time as content in Pavlovian conditioning. *Behavioral Processes* 44(2): 147–162.

Siegler, R. S., and Robinson, M. 1982. The development of numerical understanding. In H. W. Reese and L. P. Lipsitt (eds.), *Advances in child development and behavior.* Vol. 16 (pp. 242–312). New York: Academic Press.

Simon, T. J. 1999. The foundations of numerical thinking in a brain without numbers. *Trends in Cognitive Sciences* 3(10): 363–364.

Stevens, S. S. 1970. Neural events and the psychophysical law. *Science* 170(3962): 1043–1050.

Trick, L. M., and Pylyshyn, Z. W. 1994. Why are small and large numbers enumerated differently? A limited-capacity preattentive stage in vision. *Psychological Review* 101(1): 80–102.

Whalen, J., Gallistel, C. R., and Gelman, R. 1999. Non-verbal counting in humans: The psychophysics of number representation. *Psychological Science* 10: 130–137.

13　Why Animals Do Not Have Culture

David Premack and Marc D. Hauser

One of the unsatisfying things about Hamlet's monologue on human nature is that it fails to specify why we are the paragon of animals. Equally unsatisfying is the monologue of some scientists who argue that we are not so special after all. Both views are wrong-headed.

Hamlet was right in seeing us as paragons of a kind, but he failed to articulate an interesting theoretical account and failed to see the logical flaw in creating an intellectual hierarchy among animals. Specifically, why should any particular mental quality be seen as superior when every species has been equipped with a brain that was designed to solve the unique problems that emerged in its evolutionary past? Conversely, those who see nonhuman and human animals as two qualitatively similar peas in an intellectual pod have really missed what makes our own minds so different. If one can claim, without controversy, that dolphins echolocate and humans don't, why is it controversial to say that we have culture and animals don't? Sure, humans can sort of echolocate, and sure, dolphins have a sort of culture, but "sort of" is only interesting if one can specify the constraints that prevent the development of the full-blown capacity.

In this chapter we make three points:

1. Although we agree that culture must not be defined in such a way that it is uniquely human, we should use what we know about humans to formulate a theory of culture; in this sense, we adopt a position that is analogous to the debates about human language and whether other animals do or do not have anything like it.
2. We argue that the function of human culture is to clarify what people value, what they take seriously in their daily lives, what they will fight for and use to exclude or include others in their groups.
3. Based on point 2, we argue that nothing in animal behavior comes remotely close to this aspect of human culture. This does not mean that the traditions observed in animals are not interesting, but rather that we need to understand why they are so different from our own traditions.

Research on chimpanzees, whales, and dolphins suggests cultural differences among populations (see, e.g., McGrew 1992). Although most field workers acknowledge that they know little about the actual mechanisms of transmission, they are confident in their claim because neither genetic nor ecological factors can account for the variation between populations and the homogeneity within populations. Do such patterns warrant the conclusion that animals have culture, even if the behaviors are nowhere near as complicated or varied as they are in human societies? More specifically, does the notion of culture in animals help us understand its evolution in humans, or is this a misleading metaphor that might actually block important progress on this problem?

The concept of culture is one of the more elusive concepts in the social sciences (see, e.g., Sperber 1996). In harmony with many other scientists, biologists working on animal behavior define culture as any behavior that is transmitted by social learning over generations so that it becomes a population characteristic. This definition is problematic because it fails to specify the key mechanisms of cultural transmission and consequently fails to distinguish between trivial and nontrivial differences in populations. Second, a more meaningful theory of culture, and its evolution, must take into account the two key transmission mechanisms—pedagogy and imitation—in order to show why some cultural differences are trivial and others are not (see Premack 1991, Premack and Premack 2002, Tomasello 1999, on mechanisms of transmission).

Defining culture in terms of socially transmitted behaviors immediately runs into problems because some behaviors are little more than social practices, whereas others attain the status of culture. Driving on the right or left side of the road, which is beyond all doubt a socially transmitted practice, is a trivial behavior utterly lacking in social consequence (providing all conform). When on a given date and hour, Sweden changed its driving practice, Swedish culture did not change, and neither did the accident rate. On the other hand, if on the same date and hour, Sweden had discarded its Lutheran ministers, replacing them with Roman Catholic priests or Orthodox rabbis, Swedish culture would have changed dramatically. Why is the religion that Sweden practices incontrovertibly part of its culture, whereas the side of the road on which they drive is not? When culture is defined as socially transmitted behavior, this question cannot even be properly addressed. To do justice to the concept of culture, we need, not an operational definition but a theory of culture, one that will, among other things, allow trivial behaviors to be distinguished from consequential socially acquired practices. Such a theory must be built in such a way that animal culture is at least possible.

No one disputes the self-evident distinction between genetically and socially acquired behavior, but it is not a distinction that will clarify the difference between human culture and animal "traditions." Social acquisition is a secondary property of culture, neither a sufficient condition for culture, nor probably even a necessary one. If an individual acquired a culturally important idea by him- or herself, would that make the idea any less cultural?

Most work on animal culture does not do justice to the concept of social acquisition. It is our position that a theory of culture will require a clear exposition of how such acquisition mechanisms either facilitate or limit the transmission of information from generation to generation. All significant human cultural practices are transmitted by pedagogy or acquired by imitation (Galef 1992). Animals, including chimpanzees, whales, and dolphins, do not engage in pedagogy, and with the exception of vocal mimicry, evidence for motor imitation is weak as well. Imitation in chimpanzees and most other species is confined to the objects a model chooses; it does not apply to the model's actions—its motor behavior. Would-be examples of pedagogy are readily explained by something other than teaching. A possible exception is Caro's work on cheetahs and domestic cats (cited in Caro 1994; Caro and Hauser 1992), but this case also falls short of human pedagogy.

All socially acquired behaviors in chimpanzees and cetaceans appear to be of the trivial variety: holding a hand above the head when grooming, fishing for insects with a stick rather than a straw, holding hands when grooming another, soliciting sex by stripping leaves, carrying sponges on the head, lobtail fishing, beach-rubbing, wagon-wheel defense, etc. All would-be chimpanzee and cetacean cultural behaviors appear analogous to driving on one side of the road or the other. What might constitute an important social practice in the chimpanzees or cetaceans? The best candidates are likely to be found in the differences in greetings and vocal dialects. For example, it would certainly be nontrivial if groups allowed migrants in only if they immediately imitated the group's dialect or greeting gesture, or engaged in the same sort of cooperative hunting behavior. Similarly, it would certainly be nontrivial if females rejected the sexual advances of males who failed to speak their dialect. In the case of chimpanzees, those animals that hold hands when grooming might be found to deal more kindly with one another. Males that solicit sex by stripping leaves might be found to cajole females rather than force them into compliance. These consequential changes, although originating in acts that are trivial, might be found to develop slowly across generations. However, long-term observations of chimpanzees and cetaceans have so far revealed nothing of the kind. Acts that begin as trivial apparently remain trivial. They do not develop into attitudinal changes of a kind that could verge on culture.

We conclude that cetaceans and chimpanzees lack culture. This conclusion nonetheless raises many interesting questions for the future. How do humans and cetaceans differ so that while both species have social practices, only humans have cultural practices? How do cultural practices differ from mere social practices? How much of the difference between them can be explained by language?

These are difficult questions, ones that will only be answered by careful experiments investigating the psychological mechanisms that guide chimpanzee and cetacean behavior, either in the wild or in captivity.

References

Caro, T. M. 1994. *Cheetahs of the Serengeti*. Chicago: University of Chicago Press.

Caro, T. M., and Hauser, M. D. 1992. Is there teaching in nonhuman animals? *Quarterly Review of Biology* 67: 151–174.

Galef, B. G., Jr. 1992. The question of animal culture. *Human Nature* 3: 157–178.

McGrew, W. C. 1992. *Chimpanzee material culture*. Cambridge: Cambridge University Press.

Premack, D. 1991. The aesthetic basis of pedagogy. In R. H. Hoffman and D. S. Palermo (eds.), *Cognition and the symbolic processes* (pp. 303–326). Hillsdale, N.J.: Erlbaum.

Premack, A., and Premack, D. 2002. *Original intelligence—The architecture of the human mind*. New York: McGraw-Hill.

Sperber, D. 1996. *Explaining culture*. Oxford: Blackwell.

Tomasello, M. 1999. *The cultural origins of human cognition*. Cambridge, Mass.: Harvard University Press.

Contributors

Christopher Boehm
Santa Fe, New Mexico

Robert Boyd
Department of Anthropology
University of California, Los Angeles

Claude Combes
Centre de Biologie et Ecologie Tropicale
et Méditerranéenne
Université de Perpignan-CNRS
Perpignan, France

Sara Cordes
Rutgers University
New Brunswick, New Jersey

Daniel Dennett
Center for Cognitive Studies
Tufts University
Medford, Massachusetts

R. I. M. Dunbar
Evolutionary Psychology Research
Group
School of Biological Sciences
University of Liverpool
Liverpool, England

Robert A. Foley
Leverhulme Center for Human
Evolutionary Studies
University of Cambridge
Cambridge, England

Randy Gallistel
Rutgers University
New Brunswick, New Jersey

Rochel Gelman
Rutgers University
New Brunswick, New Jersey

Marc D. Hauser
Department of Psychology and Program
in Neurosciences
Harvard University
Cambridge, Massachusetts

Pierre Jaisson
Laboratoire d'Ethologie Expérimentale
et Comparée
Université Paris-CNRS
Villetaneuse, France

Stephen C. Levinson
Max Planck Institute for
Psycholinguistics
Nijmegen, The Netherlands

David Premack
Department of Psychology
University of Pennsylvania
Philadelphia, Pennsylvania

Peter Richerson
Department of Environmental Science
and Policy
University of California, Davis

Wolf Singer
Max Planck Institute for Brain Research
Frankfurt / Main, Germany

Dan Sperber
CNRS-EHESS
Institut Jean Nicod
Paris, France

Michael Tomasello
Max Planck Institute for Evolutionary
Anthropology
Leipzig, Germany